Beginner's Guide to

Understanding Geospatial Intelligence in Plain English

How Satellites and Sensors See the World We Live In

Jazper Carter

Cybersoft Publishing LLC

2026

Published by Cybersoft Publishing LLC

Part of the Serious Manager's Guide to AI and Emerging Technology series

The names of all individuals, organizations, and locations used as examples in this book are fictional. Any resemblance to real persons or entities is coincidental.

Contents

Introduction

The image stops you mid-scroll. An enormous plume of smoke, orange at the edges and charcoal gray at the center, rolls over a ridge in northern California. Underneath it, the land is already dark where the fire has passed. Above it, a second tongue of flame is racing toward a cluster of pale squares that you slowly realize are houses. The network correspondent is narrating, but you are no longer listening. You are staring at the image.

Someone made that image. A satellite, or possibly several satellites working together, captured the raw data. An analyst decided how to process it. A software pipeline chose which wavelengths to emphasize and how to map them into the red, green, and blue channels that your eyes can read. The orange color of the smoke is not accidental. Someone decided that orange would communicate danger most clearly to a general audience, and someone else reviewed that decision before the image went to air. Within minutes of that broadcast, an incident commander in a fire operations center was looking at the same image and asking a different question: how fast is the eastern flank moving, and is Route 36 still passable?

This book is about that gap. The gap between seeing a satellite image on the news and understanding what it actually is, how it was made, and what decisions ride on it every day. Closing that gap does not require a graduate degree in geography or a semester of remote sensing theory. It requires a good guide. That is what this book tries to be.

What This Book Promises

Before we go any further, here is the contract between this book and you, stated plainly.

Zero math. Not a single equation appears in these pages. The physics of how satellites work, how light bounces off surfaces, and how signals travel through the atmosphere can all be explained with everyday analogies. When the math would normally appear, we use a story or a diagram instead. If you got through high school science and never went back, you are exactly the right reader.

Zero unexplained jargon. Geospatial science has its own vocabulary, and some of it is genuinely useful. You will encounter terms like multispectral, synthetic aperture radar, and ground sample distance in the real world, and when you do, you should be able to nod along rather than feel left out. Every technical term in this book is defined in plain English the first time it appears, right where you need it. There is no glossary you are expected to flip to. The definition appears in the paragraph where the term first appears and is stated, and the term is then used naturally afterward. By the end of the book, you will have absorbed a working vocabulary without ever feeling like you were being drilled.

Every chapter anchored in a real scenario. Each chapter opens with a vignette drawn from the world of a small fictional company called Pleiades Geospatial Services. The people at Pleiades face real decisions: a wildfire moving faster than forecast, a corn grower whose crop insurance claim depends on satellite evidence, a humanitarian team trying to count displaced families from orbit, a defense analyst who needs to tell a commander whether a particular facility has been active in the past thirty days. The scenarios are fictional, but the work is real. Geospatial analysts, agricultural data scientists, disaster response coordinators, and national security professionals perform these exact tasks every day using the tools and concepts described in these pages.

A clear path to geospatial literacy. This is not a book you need to return to with a highlighter and a notepad. You can read it straight through, the way you would read any good nonfiction. By the final page, you will be able to read a satellite story in the news with genuine comprehension. You will know the questions to ask a geospatial vendor before signing a contract. You will be able to walk into a meeting where someone is showing a heat map or a change-detection layer and follow the conversation as a full participant. That is the goal.

This is not a software manual. No step-by-step instructions for ArcGIS, QGIS, or Google Earth Engine appear in this book: no code, no command lines, no configuration files. If you already know those tools, this book will deepen your conceptual understanding. If you have never opened them, this book will ensure that when you do, you are not starting from scratch.

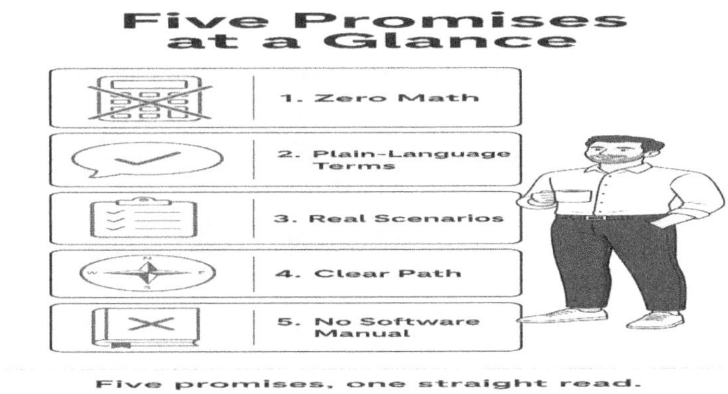

Diagram I.1: Five Promises at a Glance

Who This Book Is For

A common misconception about satellite imagery is that it belongs to a specialized guild: geographers, military analysts, remote sensing scientists, precision agriculture consultants. That is changing quickly. Today, satellite data shows up in insurance actuarial tables, city planning hearings, commodity trading terminals, investigative journalism, and NGO field reports. The tools are becoming more accessible. The vocabulary is still lagging.

If you are a data scientist who has been handed a geospatial dataset and feels like you are missing context, this book is for you. If you are a journalist who has written about satellite evidence of deforestation or arms-cache sites but secretly felt uncertain about what you were describing, this book is for you. If you manage a farm operation and your agronomist has started talking about NDVI (Normalized Difference Vegetation Index, a measure of plant health calculated from satellite data) and you want to understand the concept rather than trust the output, this book is for you. If you are a policy advisor, an urban planner, a climate analyst, or simply a curious adult who watched that wildfire image on the news and wanted to know how it was really made, this book is for you.

The one person this book is not optimized for is the working geospatial practitioner who wants deeper technical grounding. That reader should look at textbooks like Jensen's *Remote Sensing of the Environment* or Lillesand's *Remote Sensing and Image Interpretation*. This book is deliberately one level above that, written for the audience that interacts with geospatial outputs rather than producing them from scratch.

Meet the Cast: Pleiades Geospatial Services

To make the concepts concrete, this book follows a recurring group of characters through nine chapters of real work. They are employees and clients of a fictional company called Pleiades Geospatial Services, a mid-sized Earth observation analytics firm headquartered in Northern Virginia, with field partners in Iowa, California, and East Africa.

Pleiades is the kind of company that exists at an interesting intersection. It is not a satellite operator. It does not build or launch spacecraft. Instead, it buys imagery and signals data from multiple sources, applies its own analytics, and delivers processed, decision-ready intelligence to customers in agriculture, emergency management, defense, and humanitarian sectors. Think of it as the layer between raw satellite data and the people who actually need to make a decision. Companies like Pleiades are where most geospatial professionals work, and where the concepts in this book are applied every day.

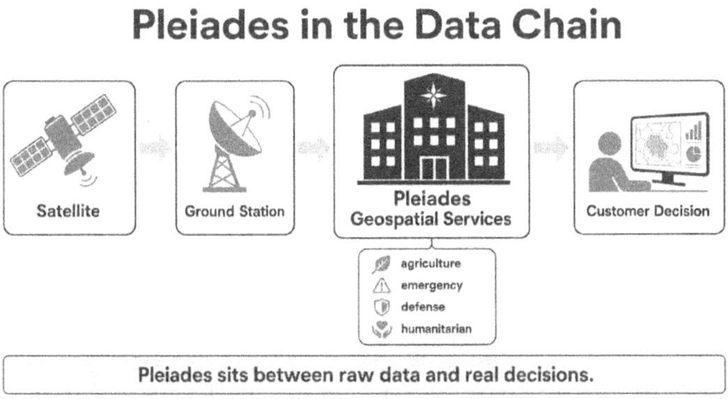

Diagram I.2: Pleiades in the Data Chain

The five people you will follow through this book represent the range of roles inside a company like this.

Daniela Reyes is Pleiades's image analytics lead. Before joining the company, she spent twelve years as a newspaper photojournalist, shooting conflict zones and natural disasters for wire services. She came to satellite imagery because she already knew how to quickly read a visual field and translate complex visual information into a story a non-specialist could understand. She is the person in the room who looks at a raw image and immediately sees the structural detail that everyone else misses. Calm under deadline pressure, decisive when the data is ambiguous, she carries the instincts of a working journalist into a field that often mistakes data density for communication.

Marcus Okonkwo joined Pleiades from a fintech firm in Lagos, where he spent five years building fraud-detection models. He is a strong data scientist who understands statistics, machine learning pipelines, and data quality. What he does not yet have is spatial intuition. He is still learning what makes geographic data different from tabular financial records, still building his sense of what a coordinate system means in practice, still occasionally surprised by the ways that physical geography shapes data patterns that a purely statistical eye would miss. Marcus is, frankly, the reader's surrogate in these pages. When Amara is explaining something that sounds obvious to the geographers in the room but opaque to everyone else, Marcus asks the question you are thinking as well. His journey from capable data scientist to capable spatial analyst runs as a quiet thread through the book.

Priya Bhatt leads Pleiades's precision agriculture practice. She grew up on her family's farm in California's Central Valley before earning an agricultural science degree and later a master's in remote sensing. She works primarily with corn and soybean growers in Iowa, sugar cane operations in Brazil, and almond orchards back in California. Priya is the most field-grounded member of the team. She has stood in rows of stressed corn at seven in the morning waiting for a drone flight, and she has sat at kitchen tables with fourth-generation farmers explaining what a satellite-derived soil moisture map actually means for their irrigation schedule. She does not traffic in abstraction. Everything she says connects directly to a field, a crop, and a decision that has to be made before the next growing window closes.

Captain Theo Lindgren is a former military GEOINT (Geospatial Intelligence) analyst who served with the United States Army for fourteen years before joining Pleiades as a senior consultant. He now works primarily with FEMA and county-level emergency managers, translating the methods and mental habits of military geospatial analysis into the civilian disaster-response context. Theo has a particular gift for pattern recognition across time. He is the person who looks at a time-series change-detection dataset and immediately sees what has moved, what has changed, and what the change implies about conditions on the ground. Disciplined and precise in his language, he is also unexpectedly collaborative. He learned in the military that a geospatial product is only as good as the decision it enables, and he has carried that customer orientation into every civilian engagement since.

Dr. Amara Wells is Pleiades's atmospheric scientist and the team's resident expert on spectral physics, the science of how light and electromagnetic radiation interact with the Earth's surface and atmosphere. She came to remote sensing from atmospheric chemistry, and she brings a deep patience for complexity to a field where people often want simpler answers than the data can honestly provide. Amara is the person who explains why the same forest looks different in an image captured in the morning versus one captured in the afternoon, why a seemingly trivial correction step actually does a lot of heavy lifting, and what the limits of any particular measurement are. She is not pedantic about it. She has spent enough time working with non-specialists to know how to teach a concept without making the listener feel as if they should have already known it.

Diagram I.3: The Pleiades Team

Beyond the Pleiades team, three external clients recur through the book. The first is a county emergency manager in a fast-growing Western county whose jurisdiction sits at the intersection of wildland and urban development. She appears in fire, flood, and heat-wave scenarios. She represents the public-safety consumer who needs fast, accurate, actionable geospatial output without having any background in how it is produced. The second is a national security customer, referred to throughout by the generic description rather than by name or agency. This character represents the intelligence community's use of satellite imagery to monitor facilities, track logistics, and build pattern-of-life assessments. The third is a program director at a humanitarian NGO that operates in East Africa and Southeast Asia. She represents the development and crisis-response sector, where satellite data increasingly support camp population estimates, agricultural shortfall forecasting, and post-disaster needs assessments.

Together, the Pleiades team and these three customers provide a comprehensive view of the landscape across the agricultural, emergency, national security, and humanitarian domains. Each domain has its own data requirements, vocabulary, and decision timelines, and the same underlying satellite data often serves all four.

A Roadmap of the Nine Chapters

This book covers geospatial intelligence from the ground up, starting with the fundamental question of how we see from space and ending with the emerging frontier of AI-driven analysis. Here is what each chapter does.

Chapter One, "Seeing from Above," establishes the conceptual foundation: why the geometry of looking down at the Earth from orbit is fundamentally different from looking at it from the ground, how altitude, angle, and resolution interact, and what it actually means to say that an image has a one-meter ground sample distance. Chapter Two, "The Electromagnetic Spectrum as a Superpower," introduces the idea that satellites can see light that human eyes cannot see and explains why this matters enormously for everything from crop health to ocean temperatures to wildfire smoke composition. Chapter Three, "How Satellites Are Built and Why It Matters," covers orbit types, sensor designs, and the commercial satellite industry, providing you with enough vocabulary to read a vendor's data sheet with ease.

Chapter Four, "From Raw Signal to Useful Image," is about the processing pipeline: what happens between the moment a sensor captures data and the moment an analyst opens the image on a screen. Chapter Five, "Reading the Land," applies those concepts to terrestrial applications: agriculture, forestry, urban development, and soil. Chapter Six, "The Dynamic Earth," examines change over time, covering how analysts detect, measure, and interpret change in deforestation, flood extent, and construction activity. Chapter Seven, "The Ocean and

Atmosphere from Orbit," takes the same conceptual toolkit offshore and upward, covering sea surface temperature, ocean color, atmospheric composition, and weather. Chapter Eight, "National Security and Geospatial Intelligence," examines how the intelligence community uses satellite data, covering GEOINT tradecraft, the open-source satellite intelligence revolution, and the ethical questions that come with near-constant overhead surveillance. Chapter Nine, "The AI Revolution in Earth Observation," closes the book by examining how machine learning is changing the speed, scale, and nature of geospatial analysis; what it can and cannot do reliably; and what a geospatial-literate professional should know before trusting automated output.

Each chapter builds on the last, but they are also designed to stand alone if you need to jump to a specific topic. If you have a pressing interest in precision agriculture, you can read Chapter Five after reading Chapters One and Two and follow it without difficulty. If your urgent need is national security or intelligence, Chapters One, Two, and Eight will give you a coherent arc. The sequence from Chapter One through Chapter Nine is the most rewarding path, but this is not a textbook with prerequisites. Start where your curiosity is strongest.

Diagram I.4: Nine-Chapter Map

A Note on What You Already Know

One last thing before we begin. You already know more about this subject than you think.

If you have ever looked at a weather radar map and understood that the red areas represent heavier rain than the green areas, you have already internalized one of the core concepts of geospatial visualization: color encodes quantity, and the encoding convention matters. If you have ever used Google Maps satellite view and noticed that recently developed areas look different from established neighborhoods, you have already practiced basic change detection. If you have ever watched a news segment that showed a before-and-after comparison of a flooded city, you have already consumed a geospatial intelligence product.

The concepts in this book will give you the vocabulary and the underlying logic to explain why those things are true. You will go from noticing to understanding. And understanding, in the long run, is considerably more useful.

Let us start at the beginning: what does it actually mean to see the Earth from above?

1 Seeing From Above

Scenario

The phone rang at 5:58 a.m. Daniela Reyes was already at her desk, coffee going cold beside her keyboard, when the call ID flashed: Kern County Emergency Management. She answered on the first ring.

The county manager's voice was clipped, controlled, the way voices get when a situation is real and moving fast. A wildfire had jumped a ridgeline overnight. Twelve hundred acres were burning as of the last ground report, but no one had eyes on the new perimeter. He needed three things before the first evacuation order could go out: which residential zones sat downwind of the fire's current head, which roads had not yet been cut by smoke or burning debris, and which water sources, whether reservoirs, tanks, or irrigation ponds, were within a reasonable helicopter bucket run from the active front. He had staff in trucks rolling toward the area, but the smoke was dense, and the terrain above the highway was steep chaparral. Ground teams could not tell him what he needed to know.

Daniela did not dispatch a field crew. She did not ask the county to wait for an aerial survey flight. She opened a workflow she had built and tested for exactly this class of problem, queried satellite imagery that had already been collected and was sitting in cloud storage, pulled a thermal infrared pass that showed heat signatures through the smoke layer, and cross-referenced road network data against burn-scar polygons detected overnight by an automated change analysis. By 7:03 a.m., the county manager had a map on his screen: a color-coded perimeter estimate, a ranked list of water sources with distances, and a traffic-light overlay on every road in the affected area. He made his evacuation call at 7:15. The chapter begins here because every concept in this book lives inside those sixty-five minutes.

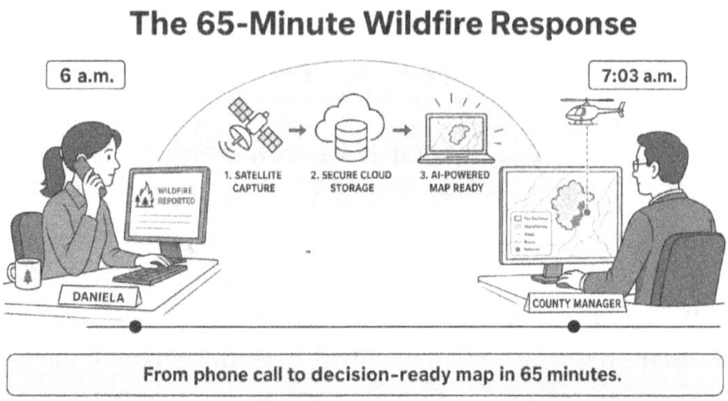

Diagram 1.1: The 65-Minute Wildfire Response

Why It Matters

If you have never thought about satellites in your workday, that story might read like a specialist's anecdote, something reserved for emergency managers and aerospace analysts. It is not. The same infrastructure that handed Daniela her map this morning is running beneath dozens of decisions you already encounter, or will encounter soon. Whether you work in agriculture, urban planning, insurance, journalism, climate research, logistics, or national security, someone in your orbit is already using geospatial intelligence to ask questions that used to require boots on the ground, months of fieldwork, or expensive aerial surveys.

The goal of this book is to make you fluent in that world without making you a satellite engineer. You do not need a background in geography, physics, or programming to read it. What you need is exactly what you are already bringing: curiosity and a willingness to think in pictures. By the last page of Chapter 9, you will be able to read a satellite-imagery story in the news and know which claims are credible and which are overstated. You will be able to walk into a conversation with a geospatial vendor, a city planner, or a defense briefer and ask sharp questions. You will understand what is technically possible, what is expensive in practice, and what is changing fast. That is geospatial literacy, and it is genuinely useful in the world we are living in right now.

Geospatial Intelligence in Plain English

Start with the word itself. "Geospatial" breaks down cleanly: "geo" from the Greek word for Earth, "spatial" from the Latin word for space or location. Put them together, and you get "relating to location on the Earth." Intelligence, in this context, does not mean a spy agency. It means processed information that is useful enough to drive a decision. Geospatial intelligence, then, is the practice of turning observations of where things are, and how they are changing, into useful information that someone can act on.

That definition sounds simple, but it carries a lot of weight. It means the raw material is always location-tagged. A photograph taken from orbit has coordinates attached to every pixel. A radar pulse bounced off a forest canopy can tell you how tall the trees are and roughly how much biomass they contain, and that measurement is anchored to a precise point on the globe. A thermal infrared scan of a city at 2 a.m. can map every rooftop, parking lot, and park by how much heat each surface has absorbed and released during the day, and every heat reading is tied to a specific address. Location is not decoration. It is the organizing principle that lets you compare observations taken at different times, from different sensors, by different organizations, and still have them talk to each other meaningfully.

Now notice what geospatial intelligence is not. It is not a map. A map is one possible output. The underlying intelligence might also take the form of a spreadsheet, a statistical model, an alert on an operations dashboard, or a set of polygons drawn around flood-risk zones that a bank uses to price insurance policies. The map is the face most people see, but the substance is the observation, analysis, and decision-making logic beneath it.

Geospatial intelligence is also not the same thing as GIS. Geographic Information Systems (GIS) refers to a class of software platforms, the most well-known of which are Esri's ArcGIS and the open-source QGIS, that store, manage, and display spatial data. GIS is one tool in the geospatial toolkit, and a very important one. But you can do a great deal of geospatial intelligence work using cloud APIs, Python spatial libraries, or web-based platforms that are not traditional GIS software at all. Thinking of the field as "GIS work" is like thinking of investigative journalism as "Microsoft Word work." The software matters, but it is not what defines the practice.

Similarly, geospatial intelligence is not the same as remote sensing, though the two terms are deeply intertwined. Remote sensing is the collection of information about an object or area from a distance, without physical contact, typically using a satellite, aircraft, or drone. It is how the raw observations get made. Geospatial intelligence is what you do with those observations once you have them. Remote sensing feeds the pipeline; geospatial intelligence is the pipeline's output. Earth observation is a broad umbrella term that encompasses both, as well as ground-based sensor networks and crowdsourced location data. Think of Earth observation as the whole orchestra, remote sensing as the strings section, and geospatial intelligence as the music those instruments produce together.

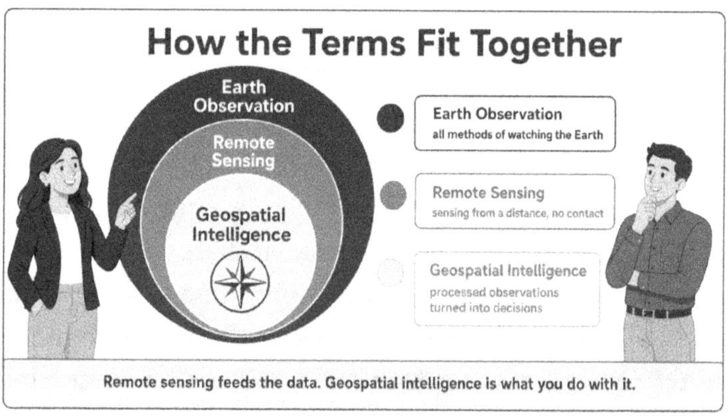

Diagram 1.2: How the Terms Fit Together

Bringing it back to Daniela's morning: the satellites passing over Kern County were doing remote sensing. They were measuring thermal energy, reflected light, and radar backscatter without any contact with the fire or the terrain. The imagery sitting in cloud storage was Earth observation data. What Daniela did with that imagery, combining it with road data, wind vectors, and topographic contours to answer a specific operational question for a specific decision-maker, was geospatial intelligence. The distinction matters because it tells you where the value is created. The value is not in the satellite. The value is in the analysis.

One more layer of terminology worth clearing up: you will sometimes hear the term GEOINT, pronounced: "jee-int." GEOINT stands for geospatial intelligence and is the formal term used by the United States intelligence community, particularly the National Geospatial-Intelligence Agency (NGA). Outside the U.S. defense context, most practitioners use "geospatial intelligence," "spatial analytics," or simply "Earth observation." In this book, you will see all of these terms. They point to the same family of ideas.

How Remote Sensing Differs From a Camera

Pull out your phone and take a picture of a park. What did you capture? Visible light, roughly the same slice of the electromagnetic spectrum that human eyes see, roughly the wavelengths between 380 and 700 nanometers. Your camera's sensor converted that slice of the spectrum into a digital image with red, green, and blue channels. The result is a faithful record of what the park looked like, in daylight, to a human eye, from that angle. That is genuinely useful and more powerful than most people realize. But it is also a very thin slice of the information available in that scene.

A remote sensing instrument does something fundamentally different. It is engineered to capture information across a much wider span of the electromagnetic spectrum, and often in ways the human eye cannot detect at all. Different parts of the spectrum reveal different things about the world. This is the single most important conceptual move in the whole field, so it is worth dwelling on.

Think of the electromagnetic spectrum as a piano keyboard. Visible light, the part your phone camera and your eyes use, occupies just a few keys in the middle of the keyboard. But the keyboard stretches far in both directions. To the right of visible light, the wavelengths get longer: first near-infrared, then shortwave infrared, then thermal infrared (which is heat), then microwaves, then radio waves. To the left of visible light, wavelengths get shorter: ultraviolet, X-ray, gamma ray. Remote sensing instruments are built to read specific keys on that keyboard, depending on what the user needs to know.

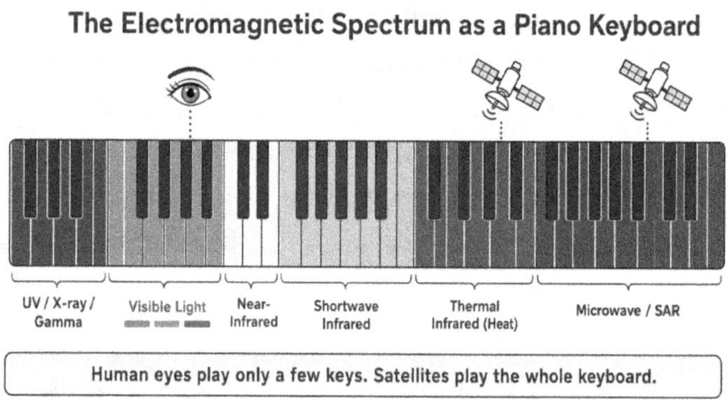

Diagram 1.3: The Electromagnetic Spectrum as a Piano Keyboard

Near-infrared (NIR) is the first key past visible red. Human eyes cannot see it. Plants reflect near-infrared light very strongly, in fact far more strongly than they reflect visible red or green. This means that a satellite sensor measuring near-infrared light can detect photosynthetically active vegetation with far more sensitivity than any visible-light camera. Healthy crops look dramatically different from stressed crops in near-infrared, even when they look identical to the naked eye. This is the physical basis for vegetation indices, which you will encounter throughout the book.

Shortwave infrared (SWIR) lets analysts distinguish between dry soil and wet soil, between rock types that look identical in visible light, and between liquid water and ice. Thermal infrared, often called TIR, measures the heat that surfaces emit rather than the light they reflect. A thermal infrared sensor flown over a city at night reveals which buildings are leaking heat (relevant for energy efficiency), which industrial facilities are running hot (relevant for environmental compliance), and, in Daniela's wildfire case, where the active fire front is burning through the smoke layer that makes visible-light imaging useless.

Synthetic aperture radar (SAR, pronounced as individual letters) operates in an entirely different regime. A SAR instrument transmits its own microwave pulses and measures the energy that bounces back. Because it generates its own illumination and because microwaves penetrate cloud cover and heavy precipitation, a SAR satellite can produce usable imagery in conditions where every optical sensor is blind. It works at night. It works through thick overcast. It works through tropical monsoon rain. SAR returns carry information about surface roughness, soil moisture, structural geometry, and subsurface conditions that no visible-light camera can see. When floods cover a delta and cloud cover blankets the scene for three weeks, SAR is often the only sensor still collecting useful data.

Diagram 1.4: What Each Sensor "Sees"

The key insight is that no single sensor tells the whole story. Each instrument is tuned to answer a particular question. An analyst building a comprehensive picture of a disaster zone, a crop field, or a coastline typically layers multiple data types: one sensor for structural damage, another for flood extent, a third for heat signatures, and a fourth for changes in vegetation health since the previous growing season. This practice of combining observations from multiple sensors and multiple spectral ranges into a single analysis is one of the defining skills in the field. It is also one of the reasons geospatial intelligence has value that no ordinary photograph can match.

LiDAR (Light Detection and Ranging) deserves a brief mention here because it often comes up alongside satellite remote sensing, even though most LiDAR is collected from aircraft rather than orbit. LiDAR fires rapid pulses of laser light and measures the time it takes for each pulse to bounce back. The result is not an image in the conventional sense. It is a dense three-dimensional point cloud, a precise elevation model of everything the laser hits: treetops, building rooftops, power lines, the ground surface beneath a forest canopy. Urban planners use airborne LiDAR to model shadow patterns and wind corridors. Forestry agencies use it to estimate timber volume. Archaeologists use it to find ancient structures buried under jungle vegetation. LiDAR is a sensing technology, not an imaging technology, and that distinction will matter when we discuss data fusion later in the book.

Why This Field Exploded in Fifteen Years

A generation ago, satellite imagery was a closely held resource. The United States and Soviet Union operated spy satellite programs during the Cold War that were classified for decades. Commercial imagery, when it became available in the late 1980s and 1990s, was expensive, often prohibitively so for anyone outside a government or large corporation. A single high-resolution image from a commercial provider could cost several thousand dollars. Analysis required specialized workstations and proprietary software. The people who did this work were a small, credentialed community.

What happened between roughly 2008 and 2023 that changed this picture so completely? Three forces converged and compounded one another.

The first force was the miniaturization of satellites. The satellite industry spent the 2000s learning that you did not need a car-sized spacecraft to do useful Earth observation. Small satellites, sometimes no bigger than a shoebox, became capable of carrying imaging sensors with resolutions that would have been classified twenty years earlier. More importantly, small satellites could be manufactured in batches, launched as secondary payloads on rockets carrying larger missions, and operated with a smaller ground infrastructure. The economics shifted. Planet Labs, founded in 2010, demonstrated that a constellation of dozens of small "CubeSats" (a standardized small-satellite format, typically about 10 centimeters on a side) could image the entire landmass of Earth every single day. Daily global coverage was not possible before this. It changed what questions were even worth asking.

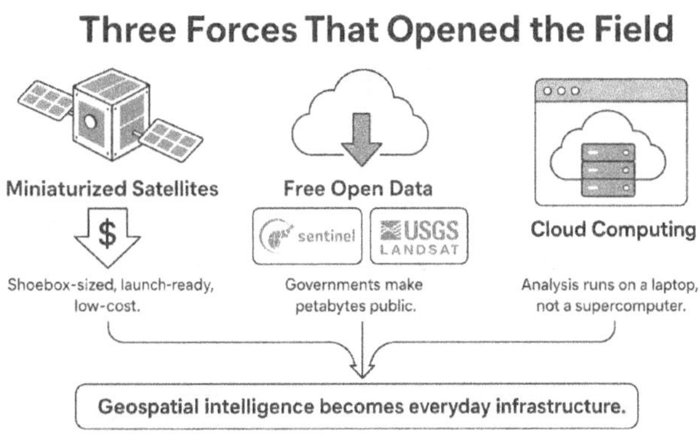

Diagram 1.5: Three Forces That Opened the Field

The second force was the release of free, open satellite data by governments. The United States Geological Survey (USGS) made the entire Landsat archive free to the public in 2008. Landsat had been collecting global imagery since 1972, meaning suddenly anyone with a browser had access to a decades-long historical record of the planet. The European Space Agency (ESA) followed with the Copernicus program, which operates the Sentinel satellite family. Sentinel-2 provides global optical imagery at 10-meter resolution every five days, free of charge. Sentinel-1 provides SAR data on a similar cadence, free of charge. Before these programs, that kind of coverage would have cost a research institution hundreds of thousands of dollars per year. Now it costs nothing.

The third force was cloud computing. The practical bottleneck with satellite data had never really been access. Even before the free data era, researchers could theoretically order imagery. The bottleneck was processing. A single Landsat scene covers about 185 kilometers by 180 kilometers, and a full global mosaic involves terabytes of data. Moving that data onto a university server and running analysis on a workstation was slow, expensive, and technically demanding. Cloud platforms changed the equation by moving the computation to where the data already lived. Google Earth Engine, launched publicly in 2010, allowed a researcher to write a few lines of code in a browser window and run an analysis across the entire Landsat and Sentinel archives, covering the whole Earth, without downloading a single file. That is not a marginal efficiency improvement. It is a category change.

These three forces did not just lower costs. They lowered the barriers to curiosity. A journalist investigating illegal deforestation could now check the satellite record without a government contract. A startup building flood insurance products for emerging markets could model risk using free imagery and cloud analysis without a satellite of their own. A graduate student in a university without a geography department could do original Earth observation research using nothing but a laptop and a browser. The community of practice grew explosively because the entry cost dropped to nearly zero.

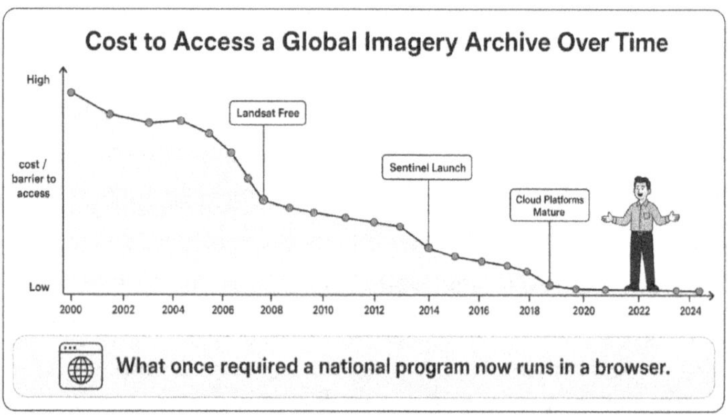

Diagram 1.6: Cost to Access a Global Imagery Archive Over Time

There is one more ingredient worth naming: the smartphone. The ubiquity of GPS-enabled phones created an enormous supply of ground-level location data, from geotagged photos and social media posts to foot-traffic signals and app-generated coordinates. This "geospatial exhaust," the location data generated as a byproduct of everyday digital life, gave analysts a new layer of information to pair with satellite observations. Combining satellite-derived building footprints with mobile-derived foot traffic patterns, for example, lets a city planner understand not just what the city looks like from above but how people actually move through it. The field today is as much about fusing these heterogeneous location data streams as it is about reading a satellite image in isolation.

Who Uses Geospatial Intelligence Today

One of the more striking aspects of geospatial intelligence in the mid-2020s is the sheer breadth of sectors that depend on it. A decade ago, the field was still strongly associated with defense and intelligence agencies. Today, those agencies remain significant users, but a much wider community surrounds them.

Farmers and agricultural companies are heavy users. Precision agriculture, the practice of managing a crop field at sub-field resolution rather than treating the whole farm as a uniform unit, depends on satellite-derived data about soil moisture, crop stress, nitrogen status, and weed pressure. An almond grower in California's Central Valley, for example, might receive weekly maps showing which blocks of trees are showing signs of heat stress, derived from thermal infrared imagery. That grower can then direct irrigation water precisely where it is needed, saving water, energy, and yield. The insurance company writing that grower's crop policy is using the same imagery to assess claims without sending an adjuster to walk every row.

The insurance sector more broadly has become a sophisticated consumer of geospatial intelligence. Catastrophe modeling for property and casualty insurers now routinely incorporates satellite-derived building footprints, roof condition assessments from very high resolution imagery, flood inundation maps, and fire-perimeter histories. When a hurricane makes landfall, a damage assessment team can begin estimating losses from satellite imagery collected within hours of the storm passing, before any adjuster reaches the ground. This speed matters enormously for claim processing and for reinsurance exposure management.

Climate scientists and environmental agencies use satellite data as the primary observational backbone of their research. No ground-based network of weather stations could cover the planet with the density or consistency that a satellite constellation achieves. Sea surface temperature, ice sheet thickness, atmospheric methane concentrations, deforestation rates, coral bleaching extent, glacier retreat, and wildfire burn area are all measured primarily from space. The long historical archives now available, Landsat going back to 1972, make it possible to quantify environmental changes over human timescales with a precision that was not achievable before.

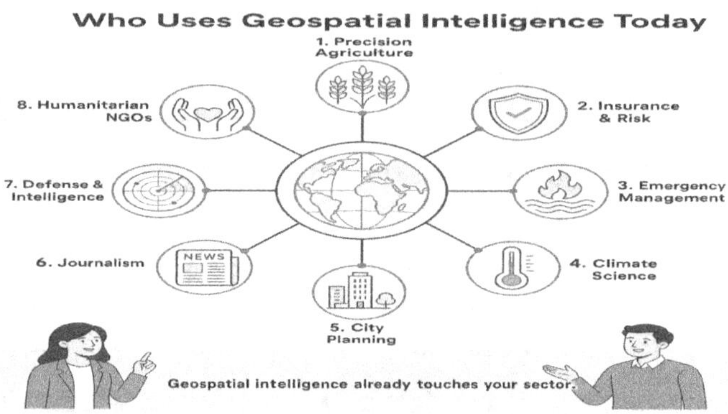

Diagram 1.7: Who Uses Geospatial Intelligence Today

City planners and urban managers use geospatial tools for everything from tracking informal housing growth in peri-urban areas to monitoring urban heat islands, mapping tree canopy cover, and planning transit corridors. A city like Phoenix, where summer ground surface temperatures can exceed 65 degrees Celsius in asphalt-heavy districts while shaded residential streets stay twenty degrees cooler, can use thermal infrared data to design targeted urban cooling interventions: strategic tree planting, cool roof programs, park placement. The analysis tells decision-makers not just where the heat is worst, but where a specific intervention would have the largest impact per dollar spent.

Journalists and open-source investigators have become increasingly sophisticated users of freely available satellite imagery. Investigative reporters have used satellite data to document illegal fishing fleets, track the construction of military facilities that governments deny exist, count vehicles at a facility to estimate activity levels, and verify or disprove claims made by governments during conflicts. The practice of using publicly available satellite imagery to verify official accounts of events, often called open-source intelligence (OSINT), has moved from a specialist niche to a recognized form of accountability journalism.

Emergency managers like Daniela's county contact depend on geospatial intelligence before, during, and after disasters. Before a hurricane season, they use satellite-derived elevation models and flood inundation models to map which neighborhoods are at highest risk. During a disaster, near-real-time satellite data helps route resources, identify isolated communities, and prioritize search and rescue. After a disaster, change-detection analysis comparing pre- and post-event imagery quantifies structural damage at a scale no field survey can match in the same timeframe.

Humanitarian organizations have incorporated satellite analysis into their operating model. When conflict displaces a population, satellite imagery can track the growth of refugee settlements, assess infrastructure destruction, and monitor agricultural disruption, keeping agencies informed when ground access is unavailable. The International Committee of the Red Cross and UN agencies such as the World Food Program now maintain in-house geospatial teams and partner with commercial providers for crisis response. In situations where the safety of field staff is a concern, satellite analysis can substitute for or supplement ground surveys in ways that were not possible a decade ago.

Shipping and logistics companies use geospatial intelligence for vessel tracking, port congestion assessment, and commodity flow analysis. An analyst monitoring global grain markets, for example, can count grain ships at Ukrainian ports using commercial satellite imagery and infer export volumes, providing a price signal independent of official statistics. The ability to observe physical economic activity from space has created a category of "alternative data" that hedge funds, commodity traders, and macroeconomic researchers use to complement conventional financial data sources.

This breadth of use is not accidental. It reflects the fact that location is a universal organizing dimension of human activity. Almost every economic, social, and environmental process has a where attached to it. Once observing the where became cheap, fast, and technically accessible, it became relevant to nearly every domain of organized decision-making.

What This Book Will and Will Not Teach

This book is a conceptual primer. That word "conceptual" is load-bearing, so it deserves explanation.

A conceptual primer teaches you how to think about a field. It builds the mental models, the vocabulary, and the judgment you need to engage with that field intelligently, whether you are reading a news article, sitting in a vendor briefing, managing a team, or evaluating a research proposal. It does not teach you button-clicking skills. It does not walk you through software interfaces step by step. It does not derive equations. Those are legitimate and valuable things to learn, but they belong in a different kind of book: a hands-on tutorial, a software manual, or a university textbook.

Here is what this book will teach you. It will teach you what electromagnetic radiation actually is and why different wavelengths reveal different properties of the Earth's surface, without requiring you to understand Maxwell's equations or radiative transfer theory. It will teach you how satellite orbits work, why some satellites are in high orbits, and others skim the atmosphere, and what those orbital choices mean for the data they collect. It will teach you how raw sensor data turns into a usable image, and what can go wrong in that process. It will teach you how analysts detect change, map objects, and classify land cover, and how machine learning has accelerated these tasks over the past few years. It will teach you how the field is applied across farming, cities, climate science, disaster response, and national security. And it will point you to free tools and datasets that let you start exploring on your own.

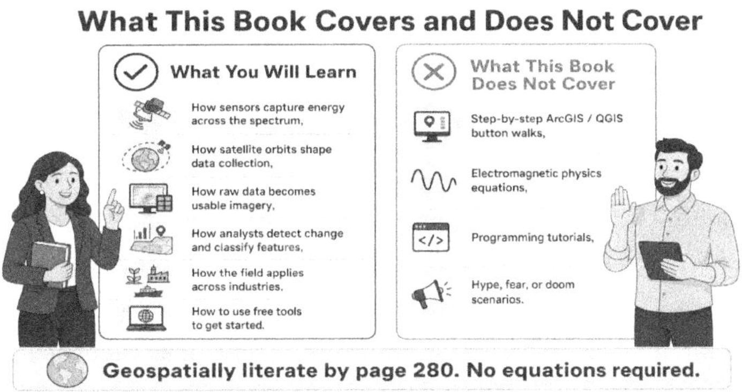

Diagram 1.8: What This Book Covers and Does Not Cover

Here is what this book will not teach you. It will not teach you to operate ArcGIS or QGIS. Both are powerful tools, and both have excellent tutorials available online and in dedicated textbooks. This book will make those tutorials easier to follow when you get to them, because you will understand what the tools are doing and why. But clicking through menus is not conceptual understanding, and a click-by-click manual would double the page count without proportionally increasing your insight. Similarly, this book will not derive electromagnetic equations. The physics of how light and microwaves interact with surfaces is genuinely interesting, and if you want the full mathematical treatment, it is available in graduate-level remote sensing textbooks. For this book, the physics will be explained at the level of the mechanism: what is happening and why it matters, without the calculus.

This book also will not teach you to code. Many of the most powerful geospatial analysis platforms today, including Google Earth Engine, Python's geopandas and rasterio libraries, and R's terra package, are programmatic. Coding fluency is enormously valuable if you want to do geospatial analysis yourself. But it is not required to understand geospatial intelligence, just as you do not need to know how to write financial software to understand corporate finance. This book will describe what those tools do and point you toward resources for learning them. The final chapter provides a curated map of where to start.

Both omissions are deliberate, and both leave you genuinely better off than if we had tried to include everything. A book that attempts to teach conceptual understanding, software operation, and programming simultaneously tends to do none of them well. This book makes a bet: if you understand the concepts deeply, you can learn the tools when you need them. The inverse is not reliably true. Many people can click through an ArcGIS workflow without having any idea what they are actually measuring or why. That is not geospatial literacy.

Meeting Pleiades Geospatial Services

Throughout this book, you will meet five professionals who work at a fictional company called Pleiades Geospatial Services. The company is named after the Pleiades star cluster, a reference geospatially readers will recognize: the Pleiades name also appears on a well-known commercial satellite constellation operated by Airbus Defense and Space. In this book, Pleiades Geospatial Services is a mid-sized analytics firm that works with clients across agriculture, emergency management, environmental monitoring, and national security. It is fictional, but everything it does reflects real workflows, real data sources, and real analytical challenges drawn from the industry.

The five core characters are:

Daniela Reyes is the image analytics lead you met at the start of this chapter. She came to geospatial work from a career as a newspaper photojournalist, and she sees satellite imagery the way a photographer sees a frame: composition, light, story. Her instincts were forged in deadline newsrooms, and she brings that same urgency and visual sharpness to every analysis. When the book needs to explain how to read an image, Daniela is usually the right character to carry the lesson.

Marcus Okonkwo is a senior data scientist who joined Pleiades from a fintech company eighteen months before the book begins. He is brilliant with models and comfortable with large datasets, but he is learning spatial data on the job. If you have ever looked at a geospatial concept and thought "wait, that doesn't quite make sense yet," Marcus is the character asking that question in the room. He is the reader's surrogate, and his questions are designed to give the author permission to slow down and explain.

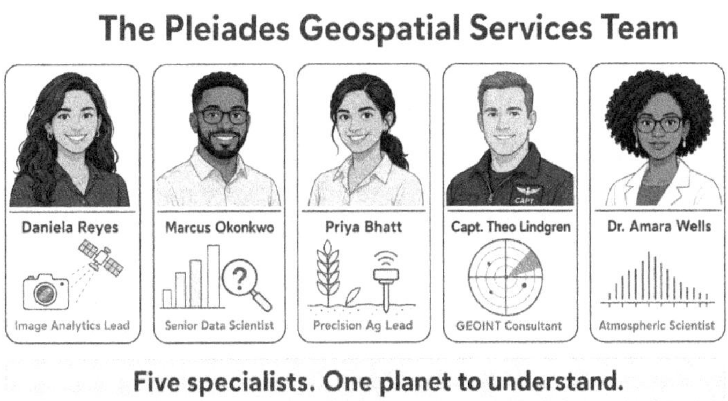

Diagram 1.9: The Pleiades Geospatial Services Team

Priya Bhatt leads the precision agriculture practice. She splits her time between the office and farm visits with clients who grow corn, soybeans, sugar cane, and almonds across four continents. Priya is practical and field-experienced, and she is often impatient with analyses that look good in a presentation but do not translate to something a farmer can act on. The agriculture chapters are primarily hers.

Captain Theo Lindgren is a former military GEOINT analyst who now consults for FEMA and county emergency managers, drawing on the skills he developed during a career analyzing imagery for military operations. Theo is disciplined, precise, and extraordinarily good at spotting patterns in data that other analysts treat as noise. He will be the primary voice in the defense, intelligence, and emergency management sections.

Dr. Amara Wells is the atmospheric scientist on the team, the one who explains why a satellite image looks the way it does, what artifacts the atmosphere introduces, and what the electromagnetic physics actually means in practice. She is a patient teacher with deep curiosity, and she has a gift for reaching for exactly the right everyday metaphor to make an abstract concept feel obvious. When the book needs to explain why the sky scatters blue light, or why thermal readings change with the time of day, Amara is at the whiteboard.

These five characters will appear and disappear across the book as the subject matter calls for them. Not every chapter requires all five. A scene with one well-chosen character telling a specific story is more useful to you than five names appearing in every paragraph without doing any work. Think of the cast as a set of guides: each one specializes in a different terrain, and the guide who appears in a given chapter is there because this is their part of the mountain.

A Map of the Next Eight Chapters

This book is organized as a journey from the physics of observation to the practice of decision-making, with industry applications woven throughout. Each chapter builds on the one before it, but most chapters can also stand alone if you already understand the foundational concepts they assume. Here is what the road ahead looks like.

Chapter 2 covers the physics of light and the electromagnetic spectrum in more depth. You got the piano keyboard version above. Chapter 2 fills in the keys: what different spectral bands actually detect, how sensor design choices shape the data you receive, and why atmospheric effects (the scattering, absorption, and emission of energy by the atmosphere itself) mean that getting from raw sensor readings to a usable, calibrated image is a non-trivial process. Dr. Amara Wells is the primary guide.

Chapter 3 gets into the hardware: satellites, orbital mechanics, and the design decisions that give different missions different capabilities. Why are some satellites in sun-synchronous orbits at 600 kilometers altitude and others in geostationary orbit at 36,000 kilometers? What is a revisit time, and why does it matter for your application? How many satellites does a constellation need to achieve daily global coverage? This chapter demystifies the hardware layer, so you understand where the data comes from before you start analyzing it.

Chapter 4 covers the journey from raw sensor data to a polished, analysis-ready image. Radiometric correction, atmospheric correction, geometric correction, orthorectification, and the concept of analysis-ready data (ARD) all live here. Think of this chapter as understanding what happens in the kitchen before food reaches the table. The transformation from photons hitting a sensor to a calibrated image with useful units involves real steps, real failure modes, and real quality implications.

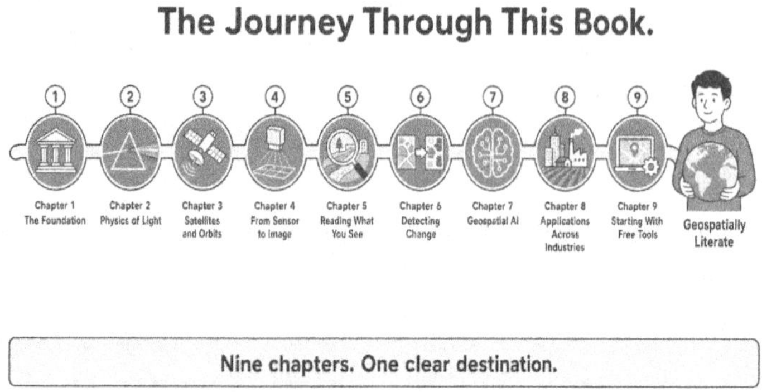

Diagram 1.10: The Journey Through This Book

Chapter 5 is about reading satellite images: what objects look like in different spectral bands, why some features that are obvious to an experienced analyst are invisible to a newcomer, and how the concepts of spatial resolution, spectral resolution, temporal resolution, and radiometric resolution all interact to determine what a given image can and cannot tell you. This chapter builds the visual vocabulary you need to interpret the imagery you will encounter in every subsequent chapter.

Chapter 6 covers change detection, arguably the most analytically powerful application of satellite data. Comparing two images of the same place taken at different times can reveal deforestation, flooding, urban growth, glacier retreat, crop failure, and damage from conflict, all without any ground visit. This chapter explains how change detection works, what its limits are, and why it is both harder and more interpretable than it first appears.

Chapter 7 covers the intersection of geospatial data and machine learning, including object detection, image classification, semantic segmentation, and time-series analysis. It explains in plain English how these methods work, where they add genuine value, and where they introduce the risk of confident-looking errors. Marcus is the primary guide here, bringing his data science background to bear on questions that matter to anyone evaluating AI-enhanced geospatial products.

Chapter 8 surveys applications across industries in depth: agriculture, urban management, environmental monitoring, humanitarian response, and financial analysis. These sections are richer than the brief industry tour in this chapter. Each application receives a comprehensive treatment of the workflow, data sources, analytical methods, and the decision it supports.

Chapter 9, the final chapter, is the practical launchpad. It surveys the free tools (Google Earth Engine, QGIS, Copernicus Browser, USGS Earth Explorer, and NASA Earthdata) and the free datasets currently available. It recommends a learning path depending on your starting point, whether you are a complete newcomer, a data scientist with no spatial background, or a domain expert looking to add geospatial literacy to your existing knowledge. It ends with the promise this book has been building toward: by this point, you have everything you need to start.

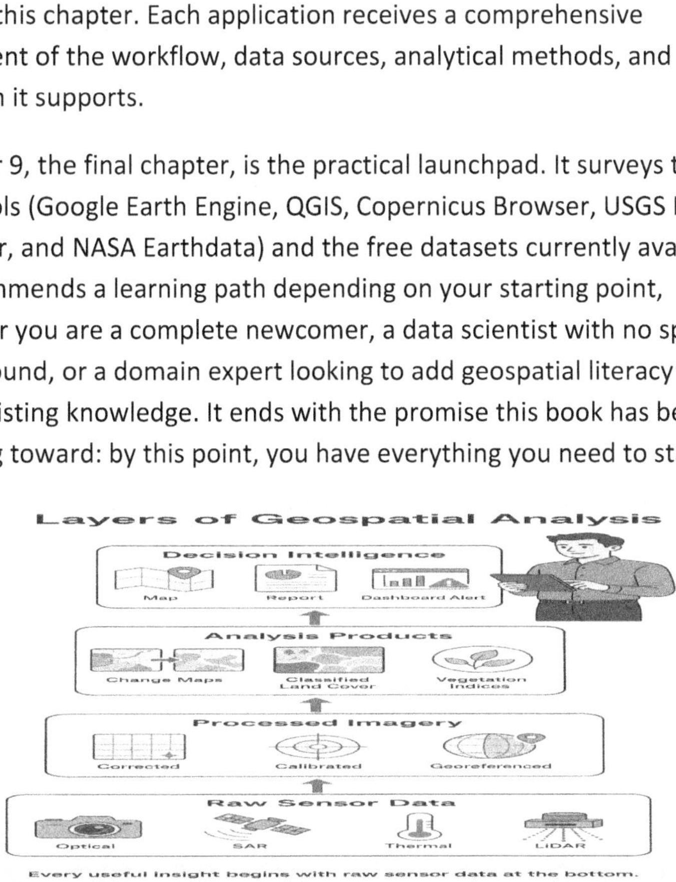

Diagram 1.11: Layers of Geospatial Analysis

One thing you will notice as you move through the book: the chapters are not arranged in order of difficulty. They are arranged in logical order. Each chapter tries to give you exactly what you need to understand the next one. If you feel the urge to skip ahead to the applications chapters, that is understandable, and you can do it. But you will get more out of Chapter 8 if you have the vocabulary from Chapters 2 through 7, just as you get more out of a music performance if you know a little about the instruments playing.

Takeaway

Sixty-five minutes. A phone call arrives; satellites already in orbit respond to a query; decades of sensor engineering, orbital mechanics, and cloud computing infrastructure converge on a specific hillside in California; and a county manager has what he needs to make a decision that will affect thousands of people. This is the world we are living in. The infrastructure exists. It is running continuously. The question is whether the people making consequential decisions in government, business, civil society, and journalism understand it well enough to use it wisely.

This book is not trying to turn you into a satellite engineer. It is trying to give you the literacy needed to participate. Geospatial literacy means you can read a news story that cites satellite evidence and know what questions to ask about resolution, revisit frequency, and sensor type. It means you can sit across from a vendor pitching a geospatial analytics product and probe the methodology rather than just looking at the output map. It means you can understand the difference between a measurement and an interpretation, and ask who made the interpretive choices and on what basis.

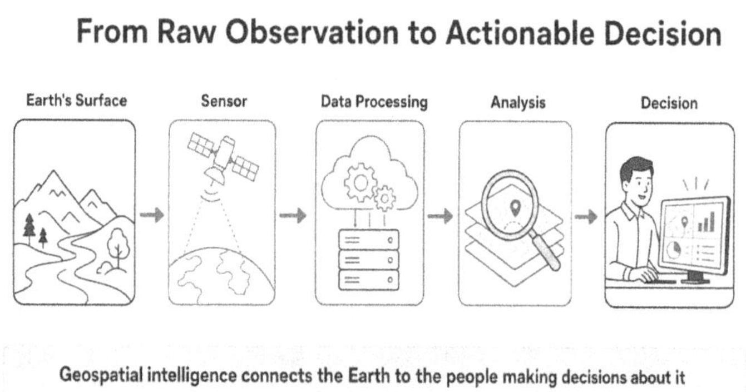

Diagram 1.12: From Raw Observation to Actionable Decision

Every concept in the chapters that follow traces back to the same pipeline you saw in Daniela's wildfire response: a sensor collects energy, a processing chain turns that energy into calibrated, georeferenced data, an analyst applies a method to extract meaning, and a decision-maker receives an output that changes what they do next. The chapters ahead cover each stage of that pipeline, with enough depth to leave you genuinely informed and enough clarity to keep you moving forward. The Earth is being watched, measured, and understood with unprecedented precision. You are about to understand how.

2 The Electromagnetic Spectrum Without the Math

Scenario

Marcus Okonkwo had been staring at spectral band combinations for three weeks, and he thought he was starting to get it. Then the Iowa grower called. Her name was Janet Hollis, and she had been farming corn in Cerro Gordo County for twenty-two years. She was polite, but she was also skeptical, the way people get when they have watched too many consultants promise things the land refused to deliver. She had done a field walk that morning. The corn looked green. The stand looked full. She did not see what the satellite was worried about.

Marcus pulled up the two images side by side on his screen and shared his desktop. On the left was the visible-color composite, assembled from the same red, green, and blue wavelengths a smartphone camera would use. The field was a perfectly ordinary green. On the right was a near-infrared false-color composite, in which healthy vegetation appears vivid red. The left half of Janet's field was bright, almost neon red. The right half was a washed-out, dull pink sliding toward grey. Marcus watched her face on the video call. She leaned forward. She was quiet for a moment.

"My eye can't see that," she said finally. It was not quite a question and not quite a statement. Marcus said that was exactly right. Her eye was built for visible light. The satellite was reading a band of energy she could not perceive at all, and in that invisible band, the right half of her field was already telling a different story. That moment- the look on her face when the invisible became visible- is the reason Chapter 2 exists.

Why It Matters

Most people come to remote sensing with the assumption that satellites are just really powerful cameras. Bigger lens, higher altitude, sharper zoom. And some of them do work roughly like that. But the overwhelming majority of what satellites actually tell us about the planet comes from wavelengths of energy the human eye cannot see at all. Understanding that fact, understanding why those invisible wavelengths exist and what they reveal, is the single conceptual key that unlocks every other chapter in this book.

You do not need to know physics to understand the electromagnetic spectrum. You do not need equations or formulas. What you need is a good mental model, a handful of everyday analogies, and the patience to sit with one new idea at a time. By the end of this chapter, you will understand why a farmer cares more about infrared than about visible green, why a wildfire analyst cares more about heat than about light, why a flood mapper can see through clouds that would blind an ordinary camera, and why every surface on Earth has what analysts call a spectral fingerprint. Once you have those four ideas locked in, everything else in this book will make sense.

The Spectrum as a Piano Keyboard

Imagine a piano. A full-size concert grand has 88 keys, spanning a range of pitches from the low rumble at the far left to the bright, crystalline high notes at the far right. Now imagine that the entire piano represents all the forms of electromagnetic energy in the universe. Gamma rays and X-rays are at the extreme left, with wavelengths so short they can pass through solid matter. Radio waves are at the extreme right, with wavelengths so long they can be measured in meters or even kilometers. Somewhere in the middle of the keyboard, spanning roughly seven keys out of the full 88, is the narrow range of wavelengths that the human eye has evolved to perceive. We call that range visible light.

Seven keys out of eighty-eight. That is how much of the electromagnetic spectrum our eyes can actually see.

The rest of the keyboard is not empty. It is full of energy that is real, physical, and measurable, energy that bounces off rocks and water and leaves and asphalt every second of every day. We do not have biological sensors for most of it. But satellites do.

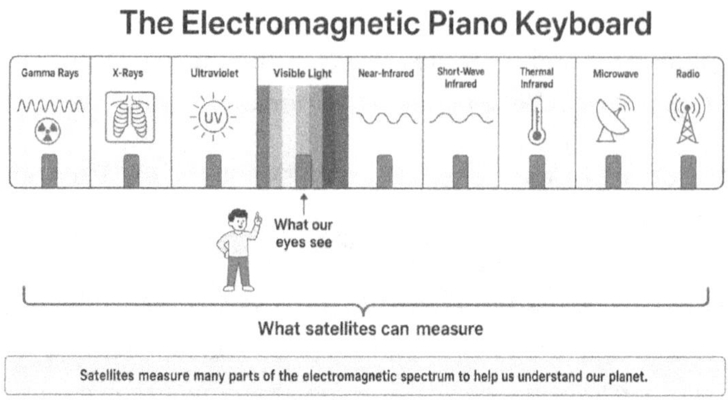

Diagram 2.1: The Electromagnetic Piano Keyboard

Let's walk along the keyboard from the visible center outward. To the left of visible light, just beyond what the eye can see, is ultraviolet (UV) radiation. You know it as the thing that causes sunburn and makes certain materials glow under a black light. Satellites that detect UV can track atmospheric ozone, monitor certain types of air pollution, and detect fluorescent materials that are otherwise invisible. You will not encounter UV sensing as often in this book as you will encounter the wavelengths to the right of visible, but it is worth knowing the key is there.

Moving to the right of visible light, you enter near-infrared (NIR). This is the first of the invisible wavelengths that remote sensing professionals use constantly, and an entire section of this chapter is devoted to it. Just past near-infrared, at slightly longer wavelengths, is short-wave infrared (SWIR). SWIR can penetrate thin clouds and smoke; it responds differently to moisture content in soil and vegetation than NIR; and it is extremely useful for distinguishing rock types in geology and for tracking wildfires through smoke. Then comes thermal infrared, which is not reflected light at all but emitted heat energy. Every object with a temperature above absolute zero emits thermal infrared radiation. Your body is glowing in thermal infrared right now, invisibly, just as a coal stove glows in visible red. Thermal sensors on satellites read temperature rather than color, which opens an entirely different set of observations.

Past thermal infrared, the keyboard enters the microwave region. These are very long wavelengths, far longer than the wavelengths of heat, and they have a property that makes them extraordinarily useful: they can travel through clouds, rain, smoke, and darkness without being absorbed or scattered. Radar sensors, including the powerful technique called Synthetic Aperture Radar (SAR), operate in the microwave region. SAR is covered in detail later in this chapter.

The key insight from the piano keyboard is this: different wavelengths reveal different things. Visible light tells you color. Near-infrared tells you plant health. Thermal tells you temperature. Radar tells you surface texture and millimeter-scale movement. A satellite that can read multiple wavelengths simultaneously is like a musician who can play chords, combining notes to reveal something neither note could express alone.

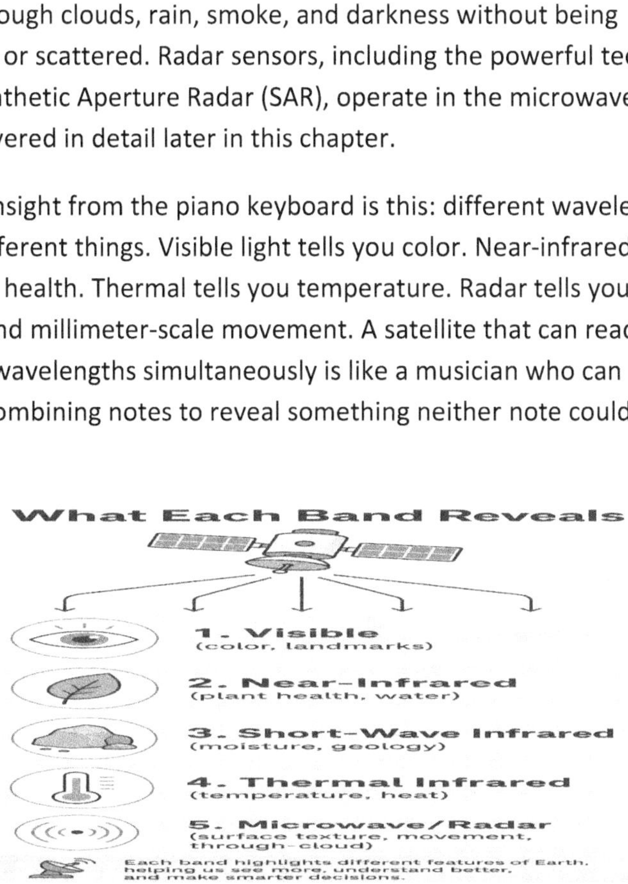

Diagram 2.2: What Each Band Reveals

One of the most common mistakes people make when they first encounter multispectral satellite data is assuming the colors in the image are real. When you see a red field in an infrared composite, that red is not what the land actually looks like. It is a deliberate choice by the analyst to assign the near-infrared channel to the red display channel, so that bright NIR reflectance appears red on screen. This is called a false-color composite, and it is a tool, not a trick. It makes the invisible visible by mapping wavelengths we cannot perceive onto colors we can. Once you understand that satellite images are often false-color composites, you can start reading them the way analysts do, by understanding what each display color represents rather than by treating the image as a photograph.

Why Infrared Is the Workhorse

If you asked a remote sensing analyst to name the single band they would keep if they could only keep one, the answer, almost universally, would be near-infrared. The reasons are practical, and they start with leaves.

A healthy green leaf does something counterintuitive. In visible light, it absorbs most of the red and blue wavelengths and reflects the green ones. That is why leaves look green to us. But in near-infrared, a healthy leaf behaves like a tiny mirror. The cellular structure of a well-hydrated, actively photosynthesizing leaf reflects near-infrared energy very strongly, bouncing back roughly half or more of the NIR energy that hits it. This is the leaf-as-mirror analogy, and it is worth holding on to because it explains much of what satellites can tell farmers and ecologists.

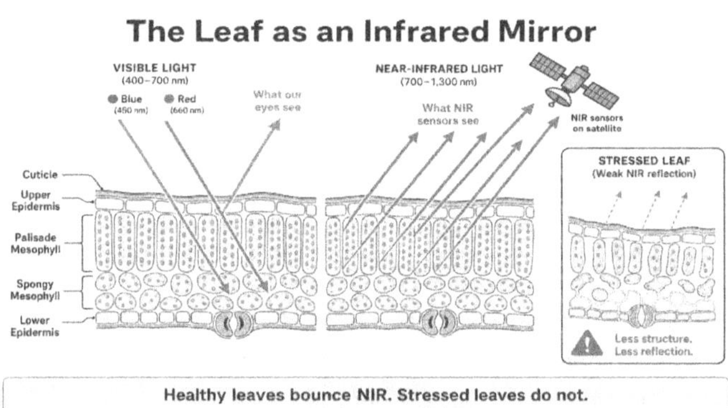

Diagram 2.3: The Leaf as an Infrared Mirror

Now think about what happens when a leaf becomes stressed. Drought stress, nutrient deficiency, early pest infestation, and fungal infection: all of these conditions alter the leaf's physical and chemical structure before damage is visible to the naked eye. The cells start to lose their turgidity. Chlorophyll levels drop. Moisture content falls. The leaf no longer bounces near-infrared as strongly. To a satellite with a NIR sensor, the change is obvious. To a person walking the field, the leaves still look green. The satellite sees trouble a week or two before the farmer's eye does.

This is not a minor advantage. In a commercial corn operation covering thousands of acres, being able to identify the stressed quarter of the field before it visibly shows damage means there is still time to intervene. A targeted application of irrigation or fertilizer to the stressed zone can save a significant portion of the yield that would otherwise be lost. The NIR band is not a curiosity. It is the financial basis of precision agriculture.

The leaf-as-mirror principle extends well beyond agriculture. Forest managers use near-infrared imaging to distinguish tree species because different species have distinct cellular structures and reflect NIR light slightly differently. Ecologists use it to track the timing of seasonal greening and browning, a measurement called phenology, which has become a key indicator of how climate change is shifting growing seasons. Ranchers and rangeland managers use it to monitor grass cover and predict grazing capacity. Water resource managers use the sharp contrast between high NIR reflectance from vegetation and very low NIR reflectance from open water to map shorelines, wetland boundaries, and irrigated fields.

Near-infrared is also well-suited to a job that seems simple but is harder than it looks: separating vegetation from non-vegetation in a scene. Pavement, rooftops, bare soil, and water all have very low NIR reflectance. Dense healthy vegetation has very high NIR reflectance. That contrast makes it easy to extract vegetation from a mixed urban or agricultural scene with a precision that visible-only imagery cannot match. This is the foundation of nearly every vegetation index you will encounter in this book.

Short-wave infrared deserves a mention here as well, because it complements NIR in important ways. Where NIR is most sensitive to leaf cell structure and chlorophyll content, SWIR is most sensitive to water content, both in leaves and in soil. When soil is wet, it absorbs SWIR strongly and appears dark in SWIR imagery. When soil is dry, it reflects more SWIR and appears brighter. The same principle applies to vegetation: drought stress shows up in SWIR before it shows up in NIR in some situations, making the combination of NIR and SWIR more powerful than either band alone. Geologists also use SWIR extensively because different mineral types exhibit distinctive reflectance patterns at these wavelengths, making SWIR imagery a standard tool for geological mapping and mineral exploration.

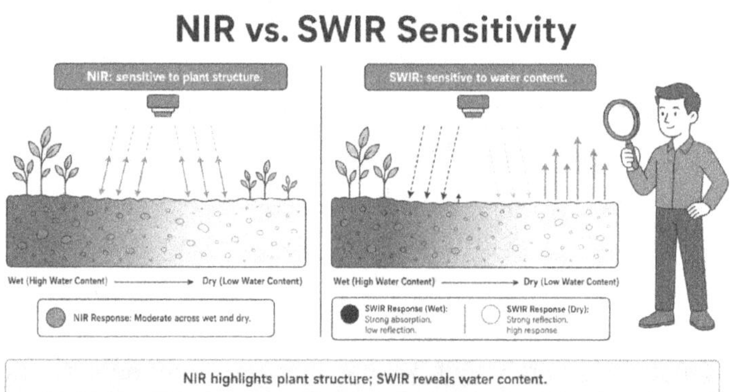

Diagram 2.4: NIR vs. SWIR Sensitivity

Thermal Imaging and What Heat Tells Us

Shift a little farther to the right along the keyboard, and you leave reflected light entirely behind. Thermal infrared sensors do not measure reflected sunlight at all. They measure emitted energy, the heat that every surface radiates based on its own temperature. This is a fundamentally different kind of observation.

You have probably seen thermal cameras in action even if you did not realize it. Home inspectors use handheld thermal cameras to find spots where insulation is missing, where air is leaking around window frames, or where water has infiltrated a wall. The camera does not illuminate the wall with any special light. It simply reads the heat the wall is already emitting and displays cold spots in blue and warm spots in yellow or white. The same principle, scaled up enormously and orbited 400 miles above the Earth, is what a thermal infrared satellite sensor does.

At the simplest level, thermal sensors let you make temperature maps of the Earth's surface. That sounds straightforward, but the applications are remarkably varied. Consider a few of them.

Urban areas absorb heat during the day from their concrete and asphalt and release it slowly at night, staying warmer than the surrounding countryside. This phenomenon is called the urban heat island effect. Thermal satellite imagery enables mapping the intensity of heat islands across an entire city, identifying which neighborhoods are hottest, and linking that thermal data to factors such as tree cover, building density, and pavement type. City planners use this information to prioritize where to plant trees or add green roofs to reduce dangerous heat exposure for residents.

Volcanoes are another obvious thermal target. A volcano building toward an eruption will often show elevated thermal emissions at its summit or along its flanks weeks or months before any surface deformation becomes visible. Satellites with thermal sensors regularly pass over active volcanic regions, and analysts track subtle changes in emitted heat as one of several early warning indicators.

Wildfires produce intense thermal signatures that cut through smoke. When a wildfire is actively burning, the hottest parts of the fire front emit heat that overwhelms the thermal sensor, allowing analysts to map not just the extent of the fire but also the location of the active fire front and the relative intensity of burning in different zones. This information goes directly to incident commanders who are making decisions about where to position firefighting resources.

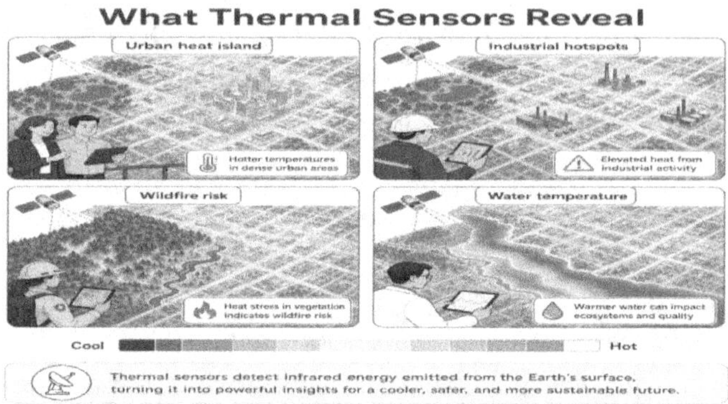

Diagram 2.5: What Thermal Sensors Reveal

There are subtler thermal applications that are equally important. Fishing fleets operating in remote ocean regions burn fuel in their engines, and those engines are warm. At night, a thermal sensor can detect the heat signatures of vessel engines, allowing maritime patrol agencies to track fishing activity in protected zones even when vessels are running without their transponders. Power plants, industrial facilities, and data centers all emit waste heat that is clearly visible in thermal imagery. Agricultural water managers use thermal sensors to detect the cooling effect of irrigation, because freshly irrigated fields are several degrees cooler than dry ones, allowing them to verify whether water is actually being applied where it is supposed to be.

The limitation of thermal sensing is that it is measuring the temperature of the surface, not the air above it. Surface temperatures fluctuate with sunlight, so most thermal analyses are done at consistent times of day, often pre-dawn when surface temperatures have had all night to equilibrate. Thermal data is also sensitive to the angle of the sun, the moisture content of the air between the satellite and the surface, and the emissivity of different surface materials (different materials radiate heat at different efficiencies). Analysts account for these factors during processing. But the core capability, the ability to turn temperature into a map, remains one of the most powerful and intuitive things a satellite can do.

Radar and SAR: Seeing Through Clouds

Every sensor discussed so far- visible, near-infrared, SWIR, thermal- shares one limitation. They all depend on clear air between the sensor and the surface. Clouds block them. Dense smoke degrades them. Rain scatters them. For significant portions of the Earth's surface and significant periods of the year, optical sensors are unable to collect data. The tropics, for example, are covered by clouds more than half the time. Much of the polar regions is in darkness for months at a stretch. A satellite that can only see in clear daylight is a useful tool. A satellite that can see through clouds, rain, smoke, and darkness is more than that.

That is what Synthetic Aperture Radar (SAR) does. The word radar itself stands for Radio Detection And Ranging. SAR is a specific form of radar built for satellites, and the way it works is fundamentally different from every sensor mentioned so far.

Think about a bat navigating in complete darkness. The bat is not seeing reflected light at all. It emits high-frequency sound pulses and listens for the echoes. The time delay between the emitted pulse and the returning echo tells the bat how far away the surface is. The strength and pattern of the echo tell the bat what the surface is made of: a hard rock wall echoes differently than a soft moth wing. The bat builds a three-dimensional map of its environment using sound, with no light required.

A SAR sensor works on the same principle, using microwave pulses instead of sound. The satellite transmits a microwave pulse toward the Earth, and then listens for the echo that bounces back. The time delay measures distance. The strength of the echo depends on the surface's roughness, its orientation relative to the sensor, and its electrical properties, which are strongly influenced by moisture content. Water is a very efficient reflector of microwave energy, so wet soil reflects SAR pulses much more strongly than dry soil. This is why SAR is an excellent flood-mapping tool: flooded areas light up in SAR imagery even when they are completely cloud-covered.

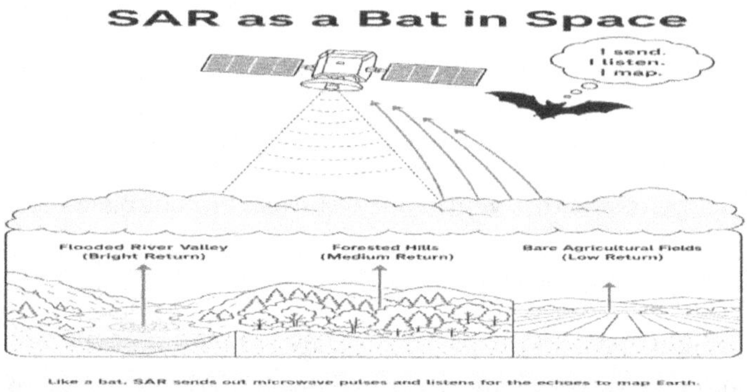

Diagram 2.6: SAR as a Bat in Space

The word "synthetic" in Synthetic Aperture Radar refers to a clever engineering trick. To get a sharp image from a radar system, you need a large antenna. But large antennas are heavy and expensive to launch. SAR gets around this by using the satellite's motion to simulate a large antenna. As the satellite flies forward, it continuously transmits and receives pulses. By combining all those pulses mathematically, the satellite's computer creates an image that has the resolution of an antenna far larger than anything that could fit on a spacecraft. This is the "synthesis": building a large virtual antenna from a small real one and applying the physics of motion.

SAR images look different from optical images. They are often displayed in shades of grey, where brightness corresponds to the strength of the radar echo rather than to color. Smooth, flat surfaces like calm water or airport runways reflect the radar pulse away from the satellite rather than back to it, so they appear very dark. Rough surfaces like forests or buildings scatter the pulse in all directions, returning some energy to the satellite, and appear brighter. Corner-shaped structures, such as building walls, that are at right angles to the satellite are extremely efficient reflectors and appear as very bright spots. Once you understand these conventions, SAR imagery becomes readable, even intuitive.

The other remarkable capability of SAR is its sensitivity to tiny movements of the Earth's surface. By comparing two SAR images of the same area taken at different times, analysts can measure ground deformation with millimeter precision. This technique is called Interferometric SAR, or InSAR. It has been used to measure the slow subsidence of cities built on soft sediment, to track the inflation of a volcanic magma chamber before an eruption, to map the extent of ground shaking after an earthquake, and to monitor the settling of roads and buildings over time. No other remote sensing technique can match InSAR for detecting millimeter-scale ground movement across large areas.

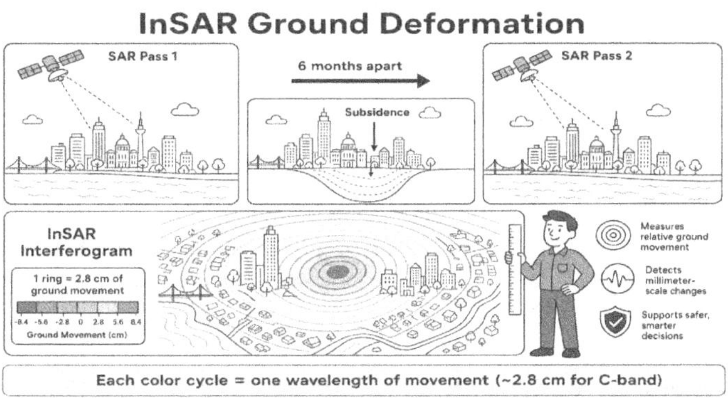

Diagram 2.7: InSAR Ground Deformation

SAR is not without its own complexity. The images require significant processing before they are usable, and interpreting them requires some experience with how different surfaces and land covers behave in the microwave region. But the core capability is transformative: SAR provides a set of eyes on the Earth that neither clouds nor darkness can close.

Spectral Signatures: Every Surface Is Different

Here is the foundational idea of remote sensing, stated as plainly as possible: every type of surface on Earth reflects and emits electromagnetic energy in its own distinctive pattern across the spectrum. That pattern is called a spectral signature. And it is the reason a satellite can look at a landscape from 400 miles up and distinguish between healthy wheat, stressed wheat, bare soil, dry soil, wet soil, asphalt, concrete, fresh snow, melting snow, clear water, and turbid water, all in the same image.

Think of a spectral signature like a fingerprint. No two people have the same fingerprint pattern, and that uniqueness is what makes fingerprints useful for identification. Similarly, no two materials have the same spectral signature, and that uniqueness is what makes spectral analysis useful for identification from space.

To understand how a spectral signature works in practice, imagine you are looking at a simple scene: a patch of healthy grass, a patch of dry, dead grass, a patch of wet soil, and a patch of dry asphalt. If you look at them in visible light, they are all different shades of green, brown, or grey. But if you plot how much energy each surface reflects at each wavelength across the whole spectrum, from visible through near-infrared through SWIR, each surface produces a different curve. The healthy grass curve rises sharply into the near-infrared. The dry grass curve stays low across the board. The wet soil curve dips strongly at the SWIR wavelengths. The asphalt curve is relatively flat and low everywhere. Those four curves are their spectral signatures.

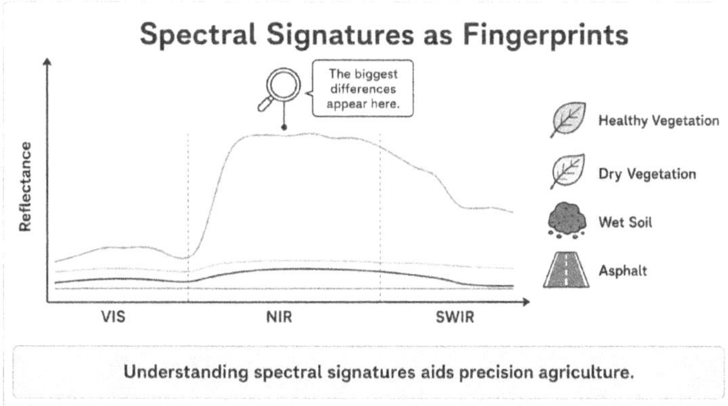

Diagram 2.8: Spectral Signatures as Fingerprints

The reason spectral signatures matter so much is that they are stable and consistent. A healthy deciduous forest in Iowa has almost the same spectral signature as a healthy deciduous forest in Romania. A spectral library built from measurements at one location can be used as a classification tool at a different location on the other side of the world. Analysts have spent decades building comprehensive spectral libraries for natural materials (vegetation types, soils, rocks, water bodies, snow, ice) and artificial materials (concrete, asphalt, steel, painted metal, certain plastics). These libraries are the reference databases that enable automated classification of satellite imagery at scale.

Environmental change shows up as a change in spectral signature. A forest being killed by bark beetles begins to show changes in its NIR reflectance months before the trees visibly turn brown. A bleaching coral reef shows a shift in its underwater spectral signature. A city neighborhood replacing green lawns with concrete and asphalt shows a shift toward the flat, low spectral profile of hard surfaces. The spectrum is, in a very real sense, the language the Earth uses to communicate its condition.

Some spectral signatures overlap, which is what makes classification challenging in real scenes. Dry bare soil and light grey concrete exhibit fairly similar spectral signatures across certain wavelength regions. Certain crop types at specific growth stages can be difficult to distinguish from one another. Shadow areas in an image suppress reflectance and can confuse classifiers. This is why analysts combine information from multiple bands rather than relying on any single wavelength, and why adding each new band to a sensor adds meaningful discriminative power. A sensor with six spectral bands can separate materials that a sensor with three bands cannot.

The concept of a spectral signature also underlies the detection of anomalies. If an analyst knows what a healthy oil palm plantation looks like spectrally, any palm that deviates from that signature stands out as an anomaly worth investigating. If an analyst knows what uncontaminated ocean water looks like spectrally, an area of water that deviates from that signature, perhaps due to an oil spill, a harmful algal bloom, or sediment discharge from a river, becomes immediately visible. The spectrum does not just tell you what something is. It tells you when something has changed.

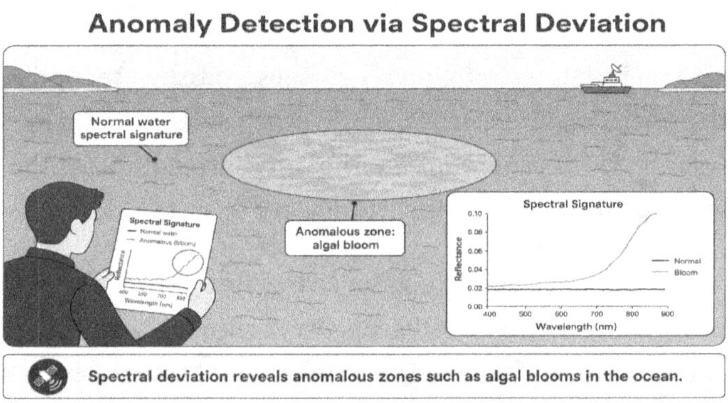

Diagram 2.9: Anomaly Detection via Spectral Deviation

Spectral Indices in Everyday Language

Analysts who work with multispectral satellite data spend a great deal of their time looking not at raw spectral bands but at derived products called spectral indices. A spectral index is a calculation, almost always a simple ratio or a normalized ratio, that combines two or more spectral bands into a single number with clear interpretive meaning. If you have ever read a news story about vegetation health measured by satellite, or a fire damage map, or a drought monitor, you have almost certainly seen a spectral index without knowing it.

Think of spectral indices as recipes. A recipe specifies proportions: one part flour, one part water, and half part fat. The ratio matters more than the absolute quantities. A spectral index works the same way: the ratio between two bands captures something meaningful that neither band expresses on its own.

The most famous spectral index in the world is NDVI, the Normalized Difference Vegetation Index. The formula is simple: take the near-infrared reflectance of a pixel, subtract the red reflectance, and divide by the sum of the two. This produces a number between -1 and 1. Bare soil and impervious surfaces cluster around zero. Sparse or stressed vegetation scores low positive values, around 0.2-0.4. Dense, healthy vegetation scores high, from 0.6 to 0.9.

Why does this ratio work? Because healthy vegetation has high NIR reflectance and low red reflectance. Those two properties push the NDVI formula toward high positive values. Stressed or sparse vegetation has lower NIR and higher red reflectance, so the ratio decreases. Non-vegetated surfaces have similar NIR and red values, so the ratio stays near zero. The subtraction and division step normalizes for overall brightness, which means NDVI is more stable across different lighting conditions than a raw band would be.

NDVI was first published in 1973 and remains one of the most widely used data products worldwide. Farmers use it to monitor field-by-field crop health. Governments use it to assess national food security. Climate scientists use long-term NDVI time series to track how vegetation cover is changing across continents. Emergency managers use it to assess drought severity. It appears in textbooks, government reports, news articles, and policy documents around the world. Understanding NDVI is not optional for anyone who works with geospatial data. It is fundamental.

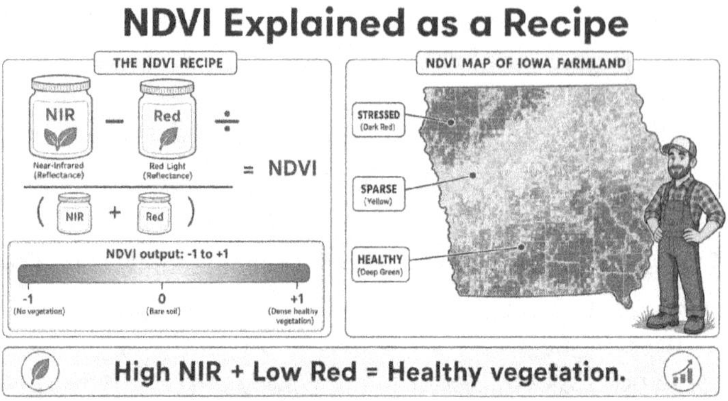

Diagram 2.10: NDVI Explained as a Recipe

The second index worth knowing is NDWI, the Normalized Difference Water Index. There are a few variations of NDWI, but the most common uses the green and NIR bands to detect surface water. Water strongly absorbs NIR and modestly reflects green, giving open water bodies a distinctive spectral ratio. Applied to a satellite image, NDWI produces high positive values over open water and negative values over land. This makes it straightforward to map rivers, lakes, floodplains, and wetlands, and to monitor changes in their extent over time. After a major flood event, analysts can compute NDWI on an image taken the day after the event and have a flood inundation map within hours. That map goes directly to emergency response coordinators.

A related index that uses SWIR instead of NIR is sometimes called the Modified Normalized Difference Water Index (MNDWI). The SWIR band is better at suppressing the signal from built-up areas, which can sometimes be confused with water in the standard NDWI. Urban analysts and flood mappers in densely built cities often prefer MNDWI for this reason.

The third index every geospatial practitioner should know is NBR, the Normalized Burn Ratio. NBR uses the NIR band and the SWIR band to detect the characteristic spectral signature of fire-damaged vegetation. Healthy vegetation has high NIR and low SWIR. Burned vegetation, bare char, and ash-covered ground have low NIR and high SWIR, the opposite pattern. NBR captures this contrast and produces a map of burn severity across a fire-affected landscape. The difference between pre-fire NBR and post-fire NBR is called the delta NBR, and it is the standard metric used by fire management agencies to produce official burn severity maps.

When you see a wildfire story in the news with a color-coded map showing areas of low, moderate, and high burn severity, you are almost certainly looking at a delta NBR product. The colors on that map are not just aesthetic. Each color class represents a specific range of spectral change that has been linked through field validation to real ecological outcomes: how much vegetation was consumed, how much soil was exposed, how long recovery is likely to take.

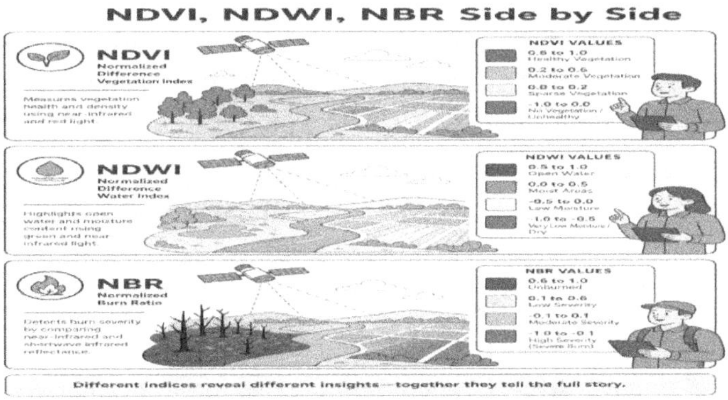

Diagram 2.11: NDVI, NDWI, NBR Side by Side

Beyond these three workhorse indices, the spectral index family is large. There are indices for estimating soil moisture, detecting chlorophyll fluorescence, separating built-up areas from bare soil, estimating snow cover, tracking algae in coastal waters, and dozens of other specialized applications. The pattern is always the same: find two or more bands where the target phenomenon creates a distinctive ratio, normalize it, and map it. The simplicity of the math belies the power of the output.

Some analysts worry that non-technical audiences will treat spectral index maps as ground truth, even though they are derived products with built-in uncertainty. The NDVI value for a field is not a direct measurement of yield or biomass. It is a proxy, shaped by sensor calibration, atmospheric conditions on the day of acquisition, soil background effects, and the specific crop growth stage. Knowing that NDVI is a ratio rather than a direct measurement helps practitioners ask better questions: Is this NDVI image from a single date, or is it a seasonal average? Was atmospheric correction applied? Which sensor collected it, and how does its calibration compare to the sensor used last year? These are the questions that distinguish a skilled analyst from someone who just learned to color a map.

Why This Chapter Unlocks the Rest

There is a specific reason this chapter appears second in the book rather than third or fifth, and it is not arbitrary. Every chapter that follows, whether it covers image interpretation, change detection, agriculture, urban planning, climate, disaster response, or national security, rests on the concepts introduced here. Once you understand the electromagnetic spectrum as a piano keyboard, you can read any satellite data story with the right mental model for which wavelengths are doing the work. Once you understand infrared as the language of plant health, every precision agriculture chapter makes sense. Once you understand thermal as a temperature map rather than a light map, every wildfire and urban heat chapter clicks into place. Once you understand SAR as a bat-pulse system operating through clouds and darkness, the flood-mapping and ground-deformation chapters follow naturally. Once you understand spectral signatures as fingerprints, the entire field of automated land cover classification becomes intuitive rather than mysterious. Once you understand spectral indices as recipes, the parade of acronyms that populate any serious geospatial report stops being intimidating.

This is why practitioners sometimes say that the spectrum is the grammar of remote sensing. You can memorize vocabulary without grammar, but you cannot build sentences. You can look at satellite images without understanding the spectrum, but you cannot reason about them. The spectrum is the progression from "I am looking at a picture" to "I understand what this observation means."

Consider how this plays out across the rest of the book. Chapter 3 covers how sensors are designed and what multispectral and hyperspectral mean as engineering terms. You will read that chapter with a clear picture of what multiple spectral bands are doing and why adding more bands increases discriminating power. Chapter 4 covers image interpretation, including the visual cues an analyst uses to identify features in a scene. Those cues are both spectral and spatial, and you will understand the spectral half completely. Chapters 5 through 9 cover application domains: agriculture, cities, climate, disaster response, security. Every single one of those chapters references specific bands or indices. You will not need to stop and decode those references. You will already have the vocabulary.

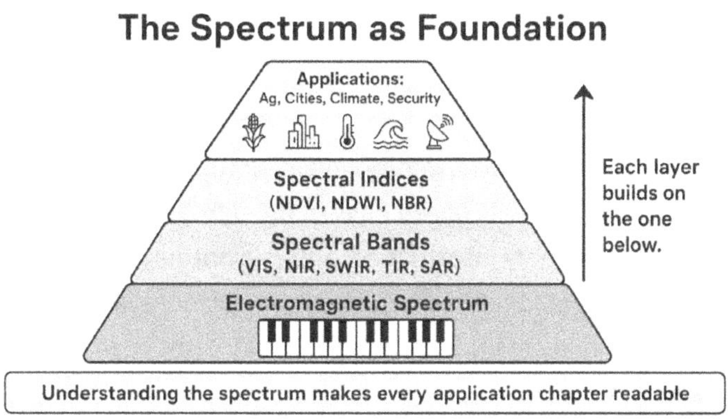

Diagram 2.12: The Spectrum as Foundation

There is also a practical benefit to this conceptual foundation outside of reading this book. When you are in a meeting, and a vendor shows you a SAR-derived flood map, you can ask whether it was an ascending or descending pass, because you understand that SAR geometry affects what surfaces produce strong returns. When you see an NDVI time series for a region, and someone claims it proves drought, you can ask what the cloud-masking protocol was, because you understand that cloudy pixels need to be excluded before the index means anything.

When a colleague mentions that thermal data was collected at 2 a.m., you can understand why, because you know that surface temperatures need to be equalized before a thermal comparison is meaningful. These are not advanced questions. They are first-principle questions that flow directly from understanding the spectrum. And asking them well is a signal of competency that opens doors.

Dr. Amara Wells, Pleiades Geospatial's atmospheric scientist, has a phrase she uses when she teaches the spectrum to new analysts: "Once you hear the whole piano, you can't unhear it." Before this chapter, when you looked at a satellite image, you were seeing seven keys out of eighty-eight, the same seven keys your eyes have always been limited to. After this chapter, you know the other eighty-one keys exist, what they sound like, and what they are good for. That is not a small thing. It is, in a very real sense, how you begin to see the world the way satellites do.

Takeaway

The electromagnetic spectrum is not a physics concept to memorize. It is a practical map of the information available to any sensor that looks at the Earth. Visible light is seven keys of an eighty-eight-key piano, and the remaining keys carry information about plant health, temperature, surface moisture, geological composition, and ground movement that no human eye can access but that modern satellites read continuously and globally.

Near-infrared is the workhorse because leaves are infrared mirrors when healthy and infrared absorbers when stressed, a distinction the eye cannot make but a sensor can. Thermal infrared turns every warm surface into a data point, mapping heat islands, volcanic activity, wildfire fronts, and irrigation patterns with a clarity that no visible camera can approach. SAR sends its own microwave pulses and reads the echoes, working through clouds and darkness the way a bat navigates a dark cave, and measuring ground deformation at millimeter precision through a technique that would seem like science fiction if it were not routine. Spectral signatures are the fingerprints that let analysts distinguish one surface type from another, not by color alone but by the full reflectance pattern across dozens of wavelengths. And spectral indices like NDVI, NDWI, and NBR are the practical recipes that turn complex spectral data into single, interpretable maps that farmers, firefighters, and flood coordinators can act on.

Carry these ideas forward. Every chapter that follows will speak the language introduced here. The spectrum is the grammar. Now you have it.

3 The Eyes In Orbit

Scenario

Theo Lindgren pulled up three images on his laptop and turned the screen toward Marcus. All three showed the same busy truck crossing on the Texas-Mexico border, captured on the same Tuesday morning in late October. To Marcus, they looked like three completely different photographs taken by three different photographers. One was grainy and pixelated, with trucks that looked like blurry gray squares. One was a little crisper but still obviously coarse. The third was so sharp you could count the axles on the tanker rigs lined up in the inspection lanes.

"Same scene, same day," Theo said, tapping each image in turn. "This one is Sentinel-2, free, from the European Space Agency. Ten-meter pixels. This one is Landsat 9, also free, from a U.S. government program run jointly by NASA and the USGS (United States Geological Survey). Fifteen-meter pixels. And this last one is commercial imagery from a provider I won't name. Fifty-centimeter pixels, ordered at about nine this morning and delivered before noon. Expensive." He let that sit for a second. "Same real world. Three very different pictures of it."

Marcus stared at the commercial image. He could see the long shadow a cab was throwing across the pavement. He could see a yellow safety vest on a customs officer near a booth. He looked at the free Sentinel image, and the same officer was a single medium-gray pixel, indistinguishable from the asphalt around them. "So why would anyone ever use the fuzzy ones?" he asked. Theo smiled. That question, he said, is the whole point of this chapter.

Why It Matters

Most people who encounter satellite imagery for the first time think the goal is to get the sharpest possible picture. The commercial image on Theo's screen looked like it had won, hands down. But the real world does not work that way, and the reason it does has nothing to do with technology. It has to do with time, money, and the specific question you are trying to answer.

If you are monitoring the health of ten million acres of farmland across three states, you cannot afford to buy commercial sub-meter imagery for every field every week. You would spend more on pixels than on seeds. But a free satellite that passes overhead every five days and delivers consistent, well-calibrated imagery at ten meters is more than sufficient. If, on the other hand, your county emergency management office needs to assess which structures were damaged in a flash flood overnight, you need the sharpest image you can get as fast as possible, and the urgency justifies the cost. The question drives the platform. The platform determines the trade-offs. And the trade-offs are not random. They follow a clear logic that, once you understand it, makes every imagery decision feel obvious rather than overwhelming.

This chapter introduces every major source of Earth observation imagery you are likely to encounter: the satellite programs funded by governments and made freely available to the world, the commercial constellations that emerged over the last decade to serve markets the government programs could not reach, the manned aircraft that still fill in important gaps, and the drones that have pushed sensing down to the field and the rooftop. It explains what a satellite image file actually looks like when it lands on a computer, why the metadata attached to that file matters as much as the pixels themselves, and where you can go right now to download your first free scene without paying a cent.

The Big Four Trade-Offs

Four fundamental tensions have shaped every satellite mission in history. Engineers call them the system trade-offs, but you do not need to be an engineer to understand them. Think of them as four levers on a mixing board, the kind a recording studio uses to balance instruments. Push one lever up and at least one of the others has to come down. No mission escapes this constraint, and no vendor pitch should be allowed to pretend otherwise.

The first lever is spatial resolution. This is the size of one pixel on the ground. A one-meter sensor captures one square meter per pixel. A thirty-meter sensor packs a thirty-by-thirty-meter patch into a single pixel. Fine pixels reveal more detail but cost more to achieve: the optics required are more precise, and the satellite is more expensive to build and launch. A fine-resolution sensor also sweeps a narrower strip per pass, so imaging a large area takes longer.

The second lever is temporal resolution, usually called revisit time. This is how often the satellite passes over the same spot. A satellite in low Earth orbit typically circles the planet about fifteen times a day. Still, any given patch of ground may be overflown only once every few days because Earth is rotating beneath it. Constellation operators solve this by launching many satellites into slightly different orbital planes so that at least one is overhead at any given time. But building and launching dozens of satellites requires capital, and customers pay for it through subscriptions and per-image fees.

The third lever is spectral resolution. Sensors can capture specific wavelengths beyond visible light, including near-infrared, shortwave infrared, and thermal infrared. High spectral resolution reveals things that ordinary cameras miss, such as water stress in a crop field or the heat front of a wildfire—but more spectral channels mean a more complex sensor with higher data volumes. Many commercial high-resolution sensors sacrifice spectral depth to deliver sharper, higher-revisit imagery. Government science missions often make the opposite bargain.

The fourth lever is cost. Free government imagery has moderate spatial and temporal resolution but is available to every researcher, farmer, journalist, and student without a purchase order. That democratizing effect has changed entire fields. Commercial imagery offers higher resolution, faster revisit times, and flexible delivery, but it requires a budget. There is no free lunch in remote sensing, but the free options are genuinely good, and the commercial options are genuinely fast.

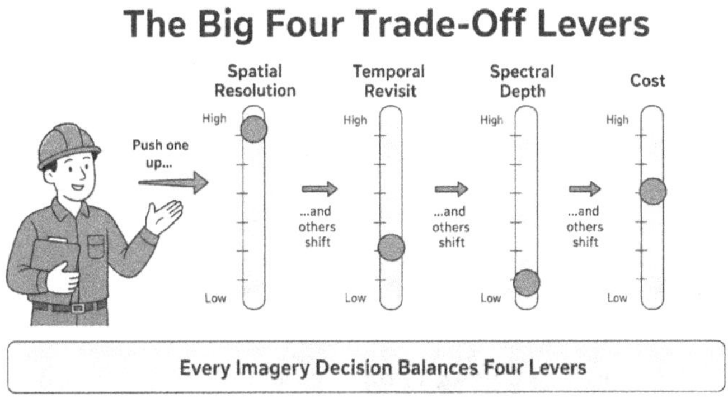

Diagram 3.1: The Big Four Trade-Off Levers

The reason Marcus's three images looked so different is precisely that each one was produced by a mission that made a different set of trade-off decisions. Sentinel-2 was designed to support agricultural monitoring, land-cover mapping, and environmental observation at continental scale. Fine enough to be useful. Free to promote maximum scientific and civic use. And engineered for consistent, repeatable measurement over time rather than for sharpest-possible point-in-time detail. Landsat has a slightly coarser pixel and a longer revisit cycle, but it has been making measurements the same way since 1972, giving analysts a fifty-year record of how the planet's surface has changed. The commercial image is optimized for a different kind of question: what is happening right now, in exquisite detail, at this exact location, and can you get it to me today? All three are right. The question determines which one wins.

Government Programs You Should Know

The most important satellite programs for everyday geospatial work are not necessarily the ones with the largest defense budgets. They are the civilian science missions that governments decided to fund and then give away for free. That decision, made first by the United States and echoed by Europe and Japan, has had enormous downstream effects on science, agriculture, disaster response, and journalism.

Landsat is the longest continuously operating land-imaging program on Earth. The first satellite in the series launched in 1972, and the program has maintained near-continuous coverage of the planet's land surface ever since. The current operational satellite, Landsat 9, launched in September 2021 and images the entire Earth every sixteen days. It is jointly managed by NASA (the National Aeronautics and Space Administration) and USGS (the United States Geological Survey), with NASA building and launching the hardware and USGS operating it and distributing the data.

The pixel size for Landsat's main instrument is thirty meters for most spectral bands and fifteen meters for a sharpened panchromatic (black-and-white) band. Thirty meters is not going to let you count axles on a truck. But it is more than fine enough to track deforestation across the Amazon, monitor the retreat of glaciers in the Himalayas, or see where an Iowa cornfield turned from healthy green to drought-stress brown over the course of a summer. The archive, critically, is free and open. In 2008, the USGS made the entire Landsat archive available at no cost to anyone with an internet connection. That single policy decision triggered an explosion of research and applications that would have been economically impossible when the scene sold data.

The archive's depth is difficult to overstate. Researchers studying coastal erosion can compare images of a barrier island from 1975 through today. Emergency managers look back at pre-fire scenes, compare them to post-fire imagery, and calculate exactly what burned. Urban planners document a city's footprint expansion across five decades. It is a historical record assembled automatically by a machine in orbit, stretching back more than fifty years.

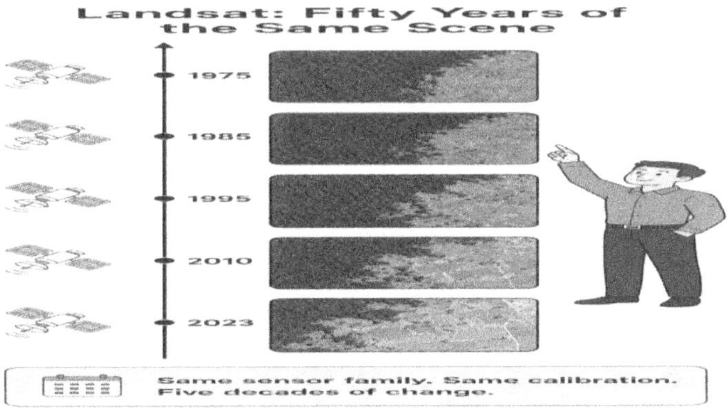

Diagram 3.2: Landsat: Fifty Years of the Same Scene

Sentinel: Europe's Free-Data Revolution

The Sentinel satellites are part of the European Union's Copernicus program, managed by ESA (the European Space Agency) in partnership with the European Environment Agency. The Copernicus program launched its first Sentinel satellite in 2014 and has since built a family of missions covering different parts of the Earth observation spectrum. Sentinel-1 carries a SAR (Synthetic Aperture Radar) instrument that can see through clouds and image the surface at night. Sentinel-2 carries a multispectral optical instrument. Sentinel-3 focuses on ocean color and sea surface temperature. Sentinel-5P monitors atmospheric composition.

Sentinel-2 is probably the workhorse that most readers will encounter first. It flies as a pair of identical satellites, Sentinel-2A and Sentinel-2B, positioned in complementary orbits so that together they can revisit any given location roughly every five days at the equator and more frequently at higher latitudes. The images are 10 meters per pixel in the key spectral bands, including red, green, blue, and near-infrared, making them well-suited for vegetation monitoring, agricultural analysis, and land-cover mapping. The data is free, open, and distributed through ESA's Copernicus Open Access Hub and through commercial cloud platforms.

The European decision to keep Copernicus data free and open was deliberate. ESA and the European Commission calculated that the economic value created by open access would far exceed the cost of managing restrictions. That calculation has proven correct. Hundreds of companies have built products on top of free Copernicus data, and researchers have done work that would have required multi-million-dollar data contracts just fifteen years earlier.

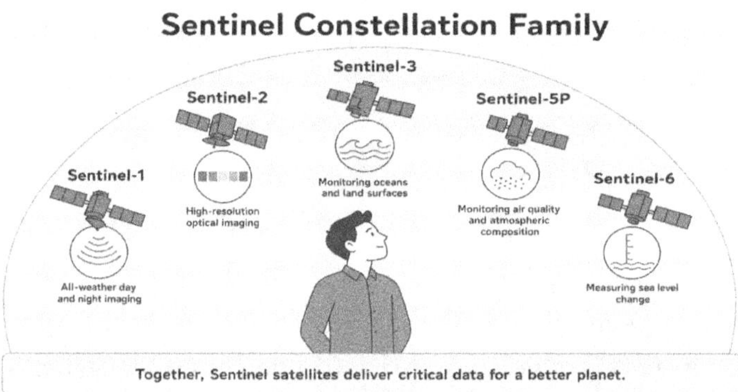

Diagram 3.3: Sentinel Constellation Family

MODIS and VIIRS: Daily Eyes on the Planet

If Landsat is the fifty-year witness and Sentinel is the five-day revisit workhorse, then MODIS (the Moderate Resolution Imaging Spectroradiometer) and its successor VIIRS (the Visible Infrared Imaging Radiometer Suite) are the Earth's daily pulse check. These instruments do not produce the kind of sharp, detailed imagery you would use to study a single field or inspect a specific building. Their pixels are hundreds of meters across. But they are designed to image the entire surface of the Earth, every single day, in dozens of spectral bands simultaneously.

MODIS flew on two NASA satellites, Terra and Aqua, launched in 1999 and 2002, respectively. Terra crosses the equator in the morning; Aqua crosses in the afternoon. Together they provided twice-daily global coverage for over two decades, capturing surface temperature, vegetation greenness, fire detections, ocean color, snow and ice extent, and dozens of other variables. MODIS data is free, and its archive is among the most widely used in all of Earth science.

VIIRS carries on MODIS's legacy aboard the Suomi NPP and JPSS (Joint Polar Satellite System) satellites with improved calibration and the same daily global reach. VIIRS fire detections are used operationally by forest management agencies worldwide to track active wildfires in near real time. Its night-lights product helps economists estimate economic activity in regions where ground-based statistics are scarce. Daily global imaging is enormously useful even at coarse spatial resolution.

Geostationary Sentinels: GOES and Himawari

There is one more class of government satellite that operates in a completely different orbital regime. Geostationary satellites sit in orbit approximately 35,800 kilometers above the equator, a distance at which their orbital speed exactly matches Earth's rotation. From the ground, a geostationary satellite appears to hang motionless in the sky. This has a profound implication: instead of sweeping a strip of ground as a low-orbit satellite does, a geostationary satellite stares continuously at the same face of the planet.

GOES stands for Geostationary Operational Environmental Satellite, and the family of GOES satellites operated by NOAA (the National Oceanic and Atmospheric Administration) has watched over North America and adjacent ocean basins since 1975. The current pair, GOES-East and GOES-West, scan the full disk of the Western Hemisphere as frequently as every ten minutes, and they can scan smaller targeted regions even more often during severe weather events. When a hurricane is approaching the Gulf Coast, meteorologists are watching GOES imagery as it updates in near real time.

Himawari is the Japan Meteorological Agency's answer to GOES, positioned to watch over the Asia-Pacific region from 35,800 kilometers above the equator. JAXA (the Japan Aerospace Exploration Agency) contributes to Japan's satellite capabilities more broadly, and Himawari-9 (the current operational satellite) delivers imagery of the Pacific basin and eastern Asia every 10 minutes across 16 spectral bands. Together, GOES and Himawari (and their counterparts operated by Europe, China, India, and South Korea) provide continuous, planet-spanning weather surveillance from orbital parking spots that never move.

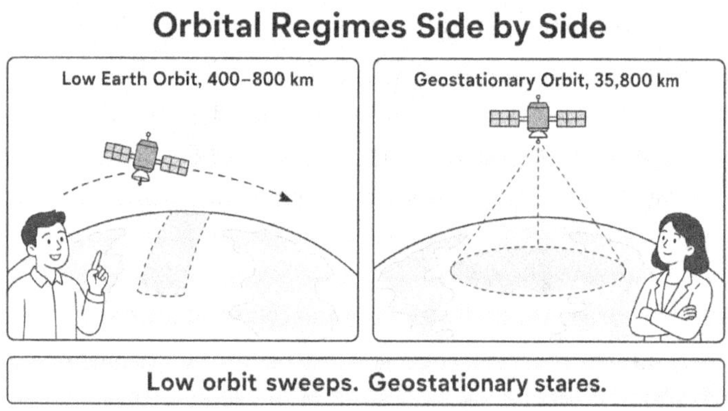

Diagram 3.4: Orbital Regimes Side by Side

Commercial Constellations and What They Unlock

For most of the first four decades of commercial satellite imagery, fine-resolution imagery came from a small number of large, expensive satellites sold through individual orders or government contracts. Imagery below one meter on the ground required a platform costing hundreds of millions of dollars to build and launch. Most researchers, planners, and businesses could not afford it.

That calculus began to change around 2013, when launch costs dropped, and sensor miniaturization advanced to the point where useful imaging capability could fit in a shoebox-sized package. A new generation of companies, backed by venture capital and staffed by engineers who had come of age during the smartphone era, bet they could build large constellations of small satellites that together would deliver more imaging capacity than the traditional high-resolution giants.

Planet Labs: The Daily Planet

Planet Labs (now Planet) is the most prominent member of that generation. Founded in 2010 by former NASA scientists, Planet has launched well over 400 small satellites. It operates a constellation large enough to image virtually every point on Earth's landmass at roughly three-meter resolution every single day. This is not the sharpest imagery on the market, but no other provider combines global coverage, daily revisit, and a model built around continuous monitoring rather than event-triggered tasking.

The practical impact has been wide. Insurance companies use Planet imagery to verify hailstorm claims without dispatching adjusters to every address. Agricultural traders use it to estimate crop area and health across growing regions in Brazil, Ukraine, and the United States. Humanitarian organizations monitor changes in refugee camp populations in hard-to-reach areas on the ground. The imagery is not free, but Planet's subscription models have brought it within reach of institutions that could never have budgeted for traditional commercial imagery.

Planet also operates a fleet of higher-resolution satellites called SkySat, which image specific targets at fifty centimeters per pixel and can capture brief video clips of the surface. Short orbital video is useful for measuring vehicle traffic flow, estimating port throughput, or tracking the movement of large equipment.

Planet's Daily Imaging Mosaic

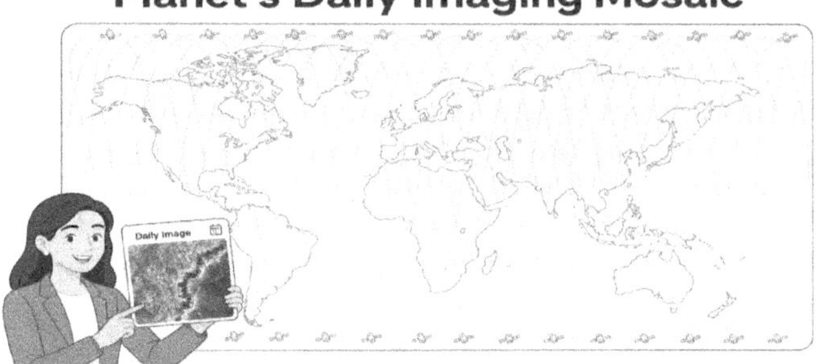

Hundreds of small satellites. One global mosaic. Every day.

Diagram 3.5: Planet's Daily Imaging Mosaic

Maxar: The Sub-Meter Standard

Maxar Technologies (formerly DigitalGlobe) operates large, high-resolution satellites that have long set the commercial optical standard. The WorldView constellation can image targets at spatial resolutions approaching thirty centimeters per pixel, fine enough to detect individual vehicles, read large printed text, or measure the dimensions of structures with reasonable accuracy.

Maxar's model is built around tasking: a customer identifies a specific target, places an order, and the satellite is programmed to image that location on its next available pass. Urgent tasking can deliver imagery within a few hours. Maxar holds long-term contracts with U.S. government agencies, and those contracts have historically enabled the company to operate satellites that would be difficult to justify on commercial revenue alone.

Capella and ICEYE: Seeing Through Clouds

Planet and Maxar both operate passive optical sensors that collect reflected sunlight, like a camera. They cannot image through clouds, and they cannot image at night. For many applications, these limits are acceptable. But for monitoring persistently cloudy regions such as the tropics during monsoon season, or for applications where waiting for a clear window is not an option, a different kind of sensor is needed.

Capella Space and ICEYE (a Finnish company) both operate constellations of small satellites carrying Synthetic Aperture Radar (SAR) instruments. A SAR instrument does not collect reflected light. Instead, it transmits a pulse of microwave energy toward the surface and measures the energy that bounces back. Because microwaves pass through clouds, rain, smoke, and darkness, a SAR satellite can produce an image of the surface regardless of weather or time of day. The resulting images look different from optical imagery (the physics of radar scattering creates a distinct texture and visual language), but they contain information that optical sensors cannot provide.

SAR is especially valuable for monitoring ships at sea, detecting flooded areas where standing water produces distinctive radar returns, tracking subsidence (slow ground sinking) in urban areas or near mining operations, and imaging through forest canopy. Both Capella and ICEYE have built constellations large enough to enable rapid revisit, with some modes imaging a target multiple times per day. Both companies serve commercial, civil government, and defense customers.

BlackSky and SkyFi: The Fast-Tasking Market

BlackSky positions itself around rapid-revisit tasking. Rather than imaging all of Earth daily or achieving sub-meter resolution, it focuses on imaging specific locations multiple times per day with very low delivery latency, often within a few minutes of collection. Automated change detection alerts customers when something changes at a monitored site, reducing the need for manual image review.

SkyFi (formerly known as Satellogic) is an Argentine-founded company that has pursued hyperspectral imaging, capturing hundreds of narrow spectral bands rather than the handful of broad bands typical of a multispectral sensor. Hyperspectral imagery is more complex to analyze but can reveal chemical and material composition information that multispectral imagery misses. SkyFi has also pursued an aggressive low-cost production model, aiming to lower the per-satellite price to the point where a large constellation is economically viable.

Together these six providers represent a market that did not meaningfully exist twenty years ago. Competition among them has driven down costs and increased capabilities, putting applications that once required defense-level budgets within reach of county planners, university researchers, and mid-sized agricultural companies.

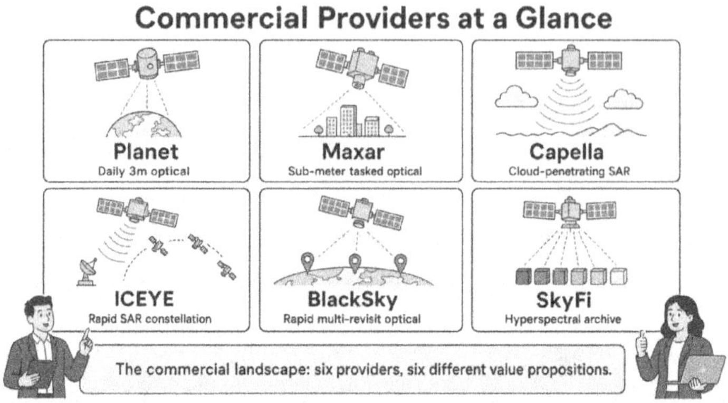

Diagram 3.6: Commercial Providers at a Glance

Aerial Surveys and Drones

Satellites are not the only platforms collecting Earth observation data from above. Long before Landsat lifted off in 1972, aerial photography from manned aircraft was the primary tool for mapping land cover and terrain. Aerial surveys remain an important part of the ecosystem, filling gaps that satellites cannot easily cover.

NAIP: The National Aerial Benchmark

NAIP stands for National Agriculture Imagery Program. It is a program run by the USDA (United States Department of Agriculture) Farm Service Agency that contracts with commercial aircraft operators to fly aerial surveys across the contiguous United States every one to three years, producing high-resolution orthorectified (geometrically corrected for terrain) imagery at one-meter resolution in four bands: red, green, blue, and near-infrared. The resulting imagery is made available for free and has become one of the most widely used baseline datasets in American GIS (Geographic Information System) work.

One-meter aerial imagery from NAIP provides a level of detail that sits between commercial satellite imagery and drone surveys. You can see individual trees, identify building footprints, trace roads and field boundaries, and assess land cover at a scale that would require expensive commercial satellite tasking to replicate from orbit. Because NAIP flights are conducted during the agricultural growing season, the imagery captures crops in their active growth phase, making it useful for crop type identification and acreage estimation. County planners, insurance adjusters, agricultural agencies, and conservation organizations all use NAIP routinely.

Outside the United States, similar programs exist at varying scales and frequencies. Many European countries maintain regular aerial photography programs for their national mapping agencies. Where coverage is sparse, commercial operators are contracted to fill gaps, particularly for infrastructure projects and environmental assessments.

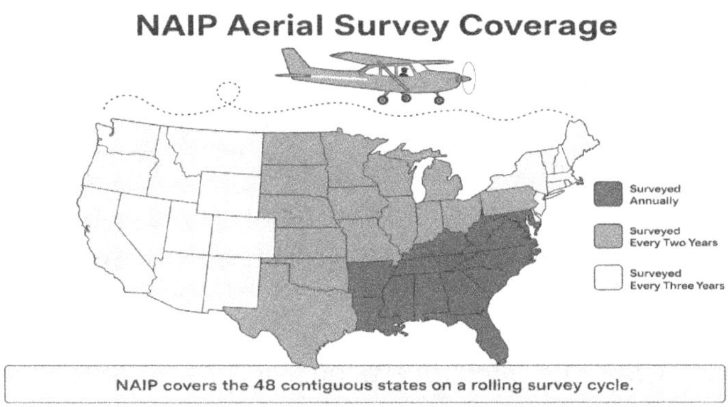

Diagram 3.7: NAIP Aerial Survey Coverage

Multispectral and Thermal Drones

The most granular data in the geospatial stack often does not come from space at all. It comes from a drone flying at one hundred meters above a cornfield, a highway, or a transformer yard. Commercial drones equipped with multispectral or thermal infrared cameras can produce imagery at resolutions of a few centimeters per pixel, revealing detail that no satellite could ever match, and they can do it on demand in a matter of hours.

Multispectral drone cameras capture the same bands as satellite multispectral sensors (red, green, blue, near-infrared, and sometimes red-edge) but at centimeter-scale resolution and over relatively small areas, typically a few hundred to a few thousand acres per flight.

Agricultural consultants routinely fly multispectral drones over precision agriculture clients' fields to produce detailed maps of vegetation health, moisture stress, or pest damage that can then be used to prescribe variable-rate applications of fertilizer or irrigation. A corn grower who identifies a stress zone early in the season may be able to intervene before yield is lost. A drone image reviewed on a Tuesday morning can inform a field operation on Wednesday.

Thermal infrared drone cameras measure surface temperature with high precision and are used in a wide range of non-agricultural applications. Utility companies fly thermal drones over power lines and substations to identify hot spots that indicate equipment about to fail, catching potential outages before they occur. Building energy auditors use thermal drone surveys to map heat loss through roofs and walls, identifying where insulation is missing or where air is leaking from the building envelope. First responders use thermal imaging to find heat signatures of people in smoke-filled buildings or dense vegetation.

The chief limitations of drone surveys are coverage and regulatory overhead. A drone can produce extraordinary imagery, but it covers only a small area per flight, and every flight requires a licensed remote pilot, battery management, and (in many jurisdictions) formal airspace authorization. Drones complement satellites; they do not replace them. They answer the question "what exactly is happening in this hundred-acre field?" Satellites answer the question "what is happening across this entire continent, and has it changed since last month?"

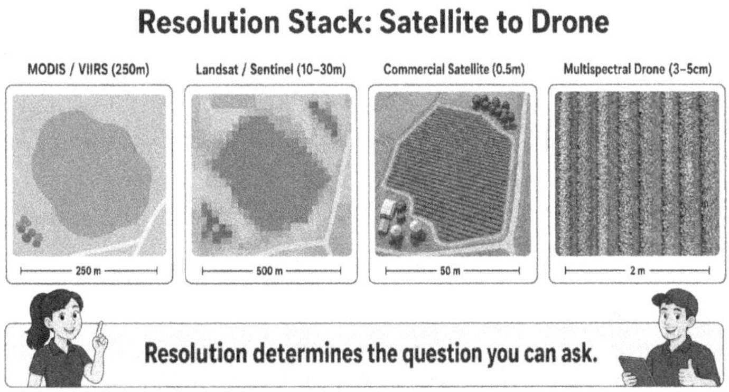

Diagram 3.8: Resolution Stack: Satellite to Drone

What a Satellite Data File Looks Like

Up to this point, the chapter has discussed platforms, missions, and trade-offs at a conceptual level. Now it is time to get slightly more concrete. When you download a satellite image, what do you actually receive? The answer is less exotic than you might expect. A satellite scene, at its core, is a digital image file, not that different in structure from the photographs on your phone, with one crucial addition: it knows where it is on the Earth.

GeoTIFF: The Standard Container

The GeoTIFF (Geographic Tagged Image File Format) is the most widely used format for distributing single satellite scenes and derived data products. The name tells you what it is: a TIFF (Tagged Image File Format). This lossless image format has been the standard for professional photography and document scanning for decades. It has now been extended to include geographic metadata embedded directly in the file header. That geographic metadata tells any compatible software exactly where on Earth each pixel is located, what coordinate system was used to project the spherical Earth onto a flat image, and the size of each pixel in real-world units.

Think of a GeoTIFF as a map that is also a photograph. An ordinary JPEG from your phone contains color data and possibly a GPS tag. Still, there is no standardized way to tell a mapping application how each pixel corresponds to a geographic location. A GeoTIFF solves this problem rigorously in a format that GIS software, remote sensing tools, and cloud platforms all understand. Open one in a compatible application, and it snaps to the correct location on the map, oriented and scaled correctly.

A Sentinel-2 or Landsat scene typically arrives as a collection of GeoTIFF files, one per spectral band. The ten-meter Sentinel-2 bands (red, green, blue, near-infrared) are four separate files. The twenty-meter bands add more. Opening the blue band gives you a grayscale image of brightness in that wavelength range. Combining red, green, and blue produces the natural-color composite that looks like a photograph. Different band combinations reveal different aspects of the landscape.

COG: The Cloud-Friendly Upgrade

GeoTIFF works well when you download an entire scene to your local disk. But satellite scenes are large. A single Sentinel-2 tile in all its spectral bands can run to a gigabyte or more. If you only want to analyze a small corner of that tile, traditional GeoTIFF requires you to transfer the whole thing before you can access any part of it.

COG (Cloud Optimized GeoTIFF) solves this problem. A COG is a GeoTIFF reorganized internally so that different parts of the image are stored in predictable locations and summary (overview) versions at coarser resolution appear at the front of the file. A client application can send a targeted HTTP range request to a cloud storage bucket and receive only the small portion it needs, rather than downloading everything. Browsing and extracting portions of large archives becomes fast even when the full dataset is petabytes in size.

The COG format has been widely adopted by major satellite data archives including NASA's Earthdata cloud, the Copernicus Data Space Ecosystem, USGS's cloud archive on AWS, and many commercial platforms. When you use Google Earth Engine to run an analysis on Landsat data, you are benefiting from COG-organized data even if you never interact with the file format directly. The convenience of Earth

Engine's interface sits atop a vast archive of COG files that are streamed to your analysis on demand.

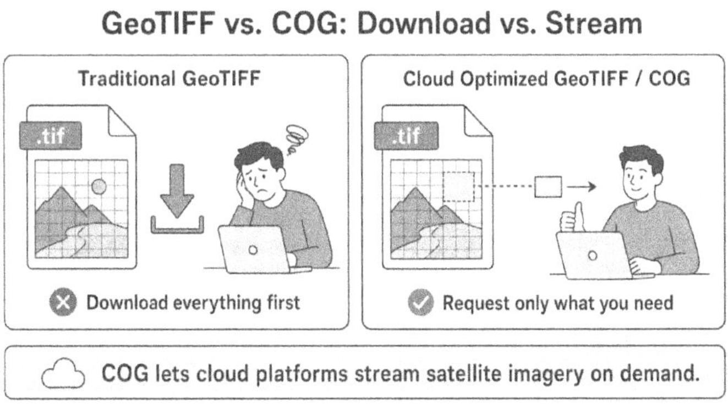

Diagram 3.9: GeoTIFF vs. COG: Download vs. Stream

Metadata: The Most Important Part

You have probably heard the term "metadata" before in contexts like email headers, music file tags, or document properties. In the world of satellite imagery, metadata is not a minor technical afterthought. It is the information that determines whether an image is useful or useless for a given purpose, and professionals who ignore it make mistakes that show up later in their analysis.

Every satellite scene comes with a metadata file or metadata embedded in the image file itself. Different providers and archives format their metadata differently, but the core fields are consistent across the industry.

The capture timestamp tells you exactly when the sensor collected the data, accurate to the second and referenced to Coordinated Universal Time (UTC). This matters because the world changes: tides rise and fall, shadows shift, crops change phenological state week to week, smoke plumes drift. Comparing two images without checking their timestamps can mean comparing incompatible states of the world.

The sensor name tells you which instrument collected the data. Different sensors have different calibration characteristics. Sentinel-2A and Sentinel-2B are nominally identical instruments with slightly different calibration coefficients. Mixing images from different sensors without accounting for these differences can introduce artifacts in time-series analyses.

The coordinate reference system (often abbreviated as CRS) tells you how the image's pixel grid maps onto the curved surface of the Earth. Every map is the result of a projection that converts three-dimensional geographic coordinates into a two-dimensional plane. Many projections exist, each suited to different extents and purposes, and an image in one projection cannot be overlaid on an image in another without a conversion step. The CRS field tells software which projection was used so that the conversion can be applied correctly.

The sun angle (also called solar zenith angle and solar azimuth angle) describes the position of the sun in the sky at the moment the image was collected. This matters for optical sensors because shadows in the image depend on sun angle. A forest canopy photographed when the sun is nearly overhead casts very small shadows; the same canopy photographed in late afternoon with a low sun angle casts long shadows that can obscure features on the ground. Comparing images taken at very different sun angles can make features appear to have changed when they have not.

Cloud cover percentage is a field that most archives populate automatically by running a cloud-masking algorithm on the image and reporting the fraction of valid (non-cloud, non-shadow) pixels. This is enormously useful for filtering large datasets. If you are looking for a clear image of a region for change-detection analysis, you can filter the archive to return only scenes with less than 10% cloud cover and skip the rest entirely. Without this field, you would have to open every scene individually and assess its usability by eye.

Processing level describes how much correction has been applied to the raw data before it reached you. Level 0 or Level 1 data is close to raw sensor output and requires additional processing before meaningful analysis. Level 2 data has been atmospherically corrected and represents true surface reflectance rather than top-of-atmosphere radiance, a distinction that matters greatly when comparing images across time or space. Most operational users want Level 2 products.

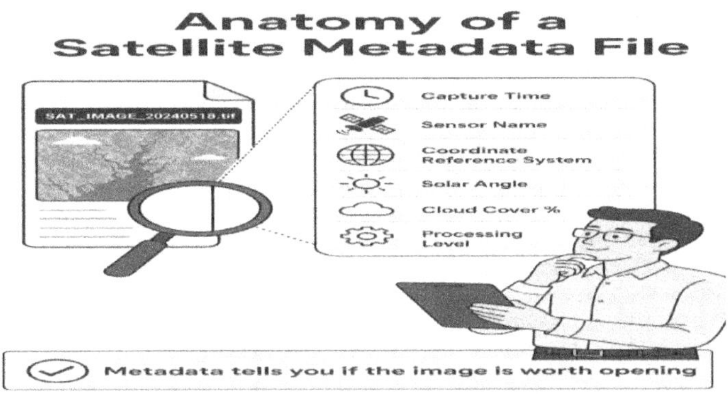

Diagram 3.10: Anatomy of a Satellite Metadata File

The rule experienced analysts follow is simple: before you look at a single pixel, read the metadata. The timestamp, sensor, projection, cloud cover, and processing level collectively indicate whether the image is appropriate for your question. An image with 70% cloud cover is not useful for vegetation analysis, even if the pixels you happen to be looking at appear clear. Comparing a Level 1 image to a Level 2 image from a different date will produce incorrect numbers. Metadata separates an analysis that answers its question from one that silently produces nonsense.

Where to Find Free Open Imagery

One of the most common misconceptions about satellite imagery is that you have to pay for it. The reality is that some of the most comprehensive and analytically useful imagery archives in the world are free, openly licensed, and accessible from a browser. The barrier to getting started is not cost. It is knowing where to look and what to do when you get there.

USGS Earth Explorer

USGS Earth Explorer (earthexplorer.usgs.gov) is the primary portal for the U.S. government's operational satellite archives. Through it, you can search and download the full Landsat archive back to 1972, current Landsat 8 and 9 data, digital elevation models, NAIP aerial imagery, historical aerial photographs back to the 1930s, and data from other sensors collected through international partnerships. The interface requires a free account and takes some initial exploration to get used to. Still, once you grasp the query structure (define an area, select a date range, filter by sensor and cloud cover), retrieving scenes is straightforward.

Copernicus Data Space

The Copernicus Data Space Ecosystem (dataspace.copernicus.eu) is the European portal for Sentinel mission data. It replaced the earlier Copernicus Open Access Hub in 2023 with a more modern interface and better cloud integration. You can browse, preview, and download Sentinel-1, Sentinel-2, Sentinel-3, Sentinel-5P, and other Copernicus products at no charge. A map viewer lets you click on a region, see available scenes with preview thumbnails and cloud cover indicators, and decide before committing to a download. The system also offers an API (Application Programming Interface) for programmatic access.

NASA Worldview

NASA Worldview (worldview.earthdata.nasa.gov) is built for fast visual browsing rather than file downloading. It functions as an interactive map you can step through in time. You can watch MODIS fire detections advance across a wildfire, see sea surface temperature shift across an El Niño event, observe dust storms blowing off the Sahara, and compare Arctic sea ice extent across years. Worldview is a useful starting point for understanding what Earth observation data looks like before you commit to downloading anything.

Google Earth Engine

Google Earth Engine (GEE) is a cloud computing platform that hosts petabytes of Earth observation data and provides both a programming environment and a graphical browser interface for running analyses without downloading anything. The public GEE data catalog includes the complete Landsat and Sentinel archives, MODIS and VIIRS products, digital elevation models, weather and climate datasets, and many other sources. The Explorer interface lets non-programmers browse the catalog, visualize layers, and understand what is available. The archive is searchable, and metadata for every dataset is well documented.

AWS Open Data

Amazon Web Services (AWS) hosts a growing catalog of public datasets through its AWS Open Data program (registry.opendata.aws), including the complete Landsat and Sentinel archives in COG format at no cost. For developers and data scientists building applications on satellite imagery, the catalog is attractive because data can be read directly into a cloud computing environment without a download step. The COG format means small geographic subsets stream efficiently without transferring entire files.

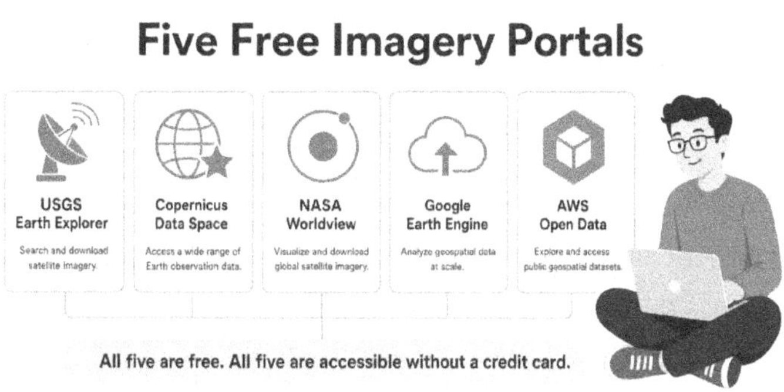

Diagram 3.11: Five Free Imagery Portals

The path from this chapter to your first downloaded scene is short. Pick an area you know: your hometown, a region you work in, a place in the news. Go to USGS Earth Explorer or the Copernicus Data Space, create a free account, draw a bounding box, select Landsat or Sentinel-2, filter by date and cloud cover, and download. The file will be a GeoTIFF with metadata embedded, ready to open in any compatible viewer. The moment you open that first scene and see a real patch of Earth in true color, correctly positioned on the globe, is one of the small pleasures of learning this field. Do not skip it.

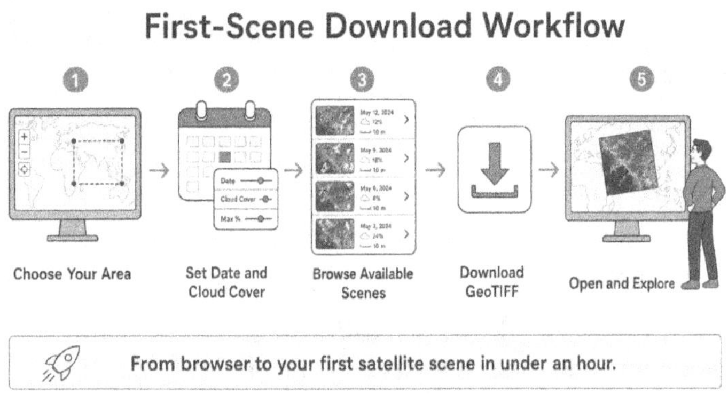

Diagram 3.12: First-Scene Download Workflow

Takeaway

The platform that produced an image determines what the image can and cannot tell you. A Landsat scene from 1985 can show you how a forest looked before a logging operation began. A Sentinel-2 scene from five days ago can show you how a crop field looks right now, in ten bands of the spectrum, at no cost. A commercial sub-meter image ordered this morning can show you whether trucks are queuing at a specific border crossing. A thermal drone survey can show you which transformer on a power line is running hot. These are four different tools answering four different questions, and none of them is universally superior.

The big four trade-offs are permanent features of the remote sensing landscape. Higher spatial resolution costs more, covers less ground per pass, and typically requires giving up something in temporal or spectral coverage. Finer temporal revisit requires more satellites or coarser resolution. Richer spectral depth requires more complex sensors and larger data volumes. Understanding these trade-offs does not require an engineering degree. It requires the same kind of practical reasoning you would apply to any procurement decision: what question am I trying to answer, what constraints am I working within, and what is the most efficient platform for that combination?

Free data has changed the field in ways that are still unfolding. The decisions made by USGS, NASA, and ESA (the European Space Agency) to open their archives at no cost created an ecosystem of research, tools, and applications that is orders of magnitude larger than what a paid-access model would have produced. The Copernicus Data Space, USGS Earth Explorer, NASA Worldview, GEE, and AWS Open Data together make it possible for a curious person with a laptop and a browser to access more satellite imagery in an afternoon than any research institution in the world could have afforded to purchase in 1990.

The chapter that follows takes the imagery you now understand how to obtain and asks the next question: what do the numbers in those pixels actually mean? You have the raw material. Now it is time to learn how to read it.

4 Reading the Picture

Scenario

Priya Bhatt pulled up the satellite map on her second monitor and frowned. The composite showed a thousand-acre corn field outside Ames, Iowa, painted in deep, alarmed reds where healthy green should have been. Her client, a third-generation grower named Dale, had walked those rows two days earlier and called the crop the best stand he had seen in years. The map disagreed loudly. Priya opened a video call and flagged the discrepancy before she had formed a theory.

Daniela Reyes joined within a minute. She said nothing at first, just studied the image the way someone reads a face. Then she asked her three questions, the ones colleagues had come to expect: What time of day was this acquired? What atmospheric correction level is tagged in the metadata? Has anyone done a geometry check against the field boundaries? Priya scrolled to the metadata panel and read back the answers. The satellite had passed at 10:47 in the morning, local time. The product was tagged Level 1C, meaning it had been corrected for sensor geometry but not for atmospheric effects. The field boundary overlay was off by about forty meters to the northwest.

The third answer was the one that cracked it. A forty-meter offset put the apparent stressed zone precisely where an isolated cumulus cloud had been drifting, casting a shadow across the northwest corner at capture time. The pixel values in that shadow band were low across all visible and near-infrared channels, and the analysis algorithm interpreted them as stressed biomass rather than temporary darkness. Daniela marked the shadow perimeter on screen with a soft blue polygon and said, "Cloud shadow artifact. The field is fine." Dale's crop was healthy. The satellite had not lied, but it had needed someone to read it correctly.

Why It Matters

That scene happens dozens of times a day in imagery analytics shops around the world. The satellite data is honest—the sensor records exactly what it measured. But a measurement is not yet an interpretation, and interpretation is where most mistakes live. Clouds, shadows, spectral compositing choices, uncorrected atmospheric haze, poorly labeled training pixels, and overeager object detectors all shape what a finished map claims to show. A reader who understands those layers is harder to fool and faster to spot a problem. This chapter gives you that understanding. You do not need to write a line of code. You need to know what questions Daniela asked and why they were the right ones.

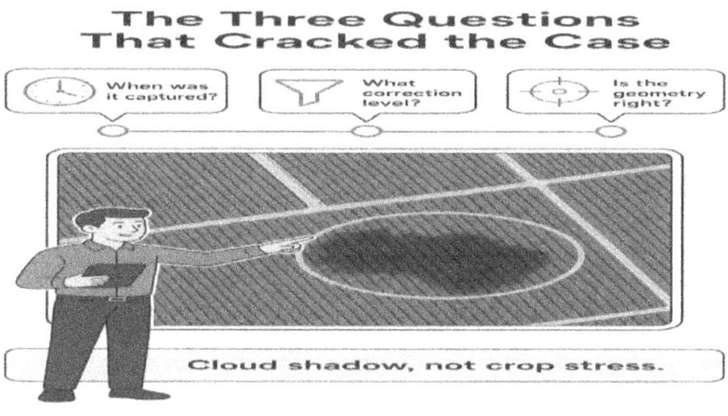

Diagram 4.1: The Three Questions That Cracked the Case

True Color, False Color, and Not Photographs

When a news outlet publishes a satellite image of a wildfire, a flood, or a missile site, the image usually looks like a photograph. It has blue sky, brown earth, and green vegetation arranged in roughly the colors you expect. But almost nothing about that image was captured the way a phone camera captures a selfie. Understanding the difference between a true-color composite and a false-color composite is the first step toward reading any satellite image with confidence.

A digital camera on your phone captures three channels of light: red, green, and blue. It blends those three channels into a single image that approximates what your eye would see. A multispectral satellite sensor operates on a similar principle but is very different in practice. Sensors like those on Landsat or Sentinel-2 record eight, twelve, or more distinct bands, each a narrow slice of the electromagnetic spectrum. Some of those bands correspond to visible colors. Others capture near-infrared, shortwave infrared, or thermal energy that your eye cannot see at all.

A true-color composite is built by mapping three of those bands directly onto the red, green, and blue display channels of your screen. When a satellite captures a red band, a green band, and a blue band, and an analyst displays them as red, green, and blue respectively, the resulting image looks close to what your naked eye would see from an airplane. Coastlines look blue. Healthy forests look green. Snow looks white. That familiarity is useful for communication. It is also, for analytical purposes, a waste of information. The sensor collected a dozen spectral channels, and the analyst is showing only three.

A false-color composite uses the same three-band display slot but fills them with different channels. The most common swap in agriculture and ecology is the color infrared composite, which maps the near-infrared band to the red display channel, the red band to the green channel, and the green band to the blue channel. The result looks strange to an untrained eye. Healthy vegetation, which reflects near-infrared light intensely, blazes bright red. Water, which absorbs near-infrared radiation, appears nearly black. Urban pavement sits in pale cyan. The image is "wrong" by the standard of human vision, but indispensable by the standard of distinguishing healthy plants from stressed ones.

Other composite combinations serve other purposes. Pairing shortwave infrared bands with near-infrared makes burn scars appear deep brown, while unburned forest remains bright. A geological composite tuned for clay minerals helps mining geologists spot potential ore deposits. A composite that emphasizes thermal emission can distinguish hot industrial facilities from cool residential areas. Each combination is a deliberate choice to highlight a specific physical phenomenon by routing non-visible light into the visible display channels.

The implication for anyone reading satellite imagery in the news is straightforward: the colors are not necessarily real. They are choices made by an analyst, a publication, or an algorithm. A red blob over a rainforest does not mean fire unless the caption confirms it means fire. An image in which vegetation appears fluorescent pink is almost certainly a false-color composite chosen for scientific utility rather than artistic effect. Asking "what bands make up this image?" is as basic as asking "what scale is this map?" and equally important.

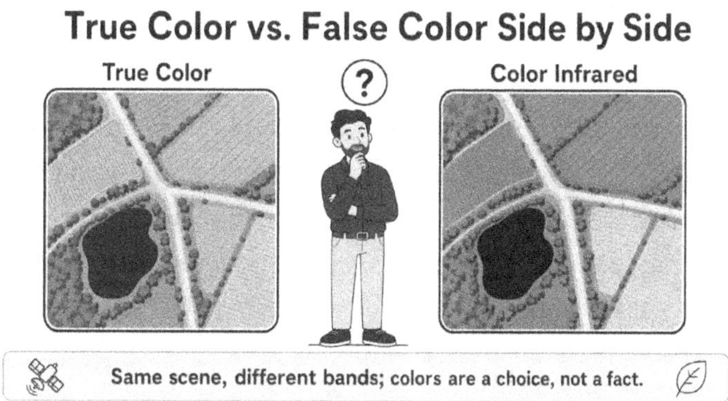

Diagram 4.2: True Color vs. False Color Side by Side

The term "band" is worth locking in here. A band is simply one channel of the sensor's recording, corresponding to a specific range of wavelengths. When you hear someone say "a 12-band multispectral satellite," they mean the sensor recorded the scene in twelve separate spectral slices simultaneously. Every pixel in the image has twelve values instead of three. An analyst can combine any three of those twelve values into a visual composite, which means there are hundreds of possible views of the same location, each revealing different information. The art and science of satellite image analysis begins with choosing the right combination for the question you are trying to answer.

One more term worth knowing is "band ratio." Sometimes an analyst does not display raw band values at all. Instead, they divide one band by another to create a derived index. The most famous example is the Normalized Difference Vegetation Index, or NDVI, which compares the near-infrared band to the red band to produce a single number that correlates closely with the amount of living green plant matter covering a surface. The math is simple. The usefulness is enormous. NDVI is not a false-color composite in the traditional sense because it does not map three bands onto red, green, and blue display channels. It is a single derived layer typically displayed as a color ramp from brown (low, no vegetation) to bright green (high, dense vegetation). When a news story says "the satellite showed the forest was thinning," NDVI or a close relative was almost certainly the tool they used.

The Four Artifacts Every Analyst Spots

An artifact in satellite imagery is anything in the image that came from the acquisition process itself rather than from the real world. The four most common artifacts are clouds and their shadows, sensor stripes, mosaic seams, and saturated pixels. Recognizing them at a glance is one of the first skills a working analyst develops, because failing to recognize them produces maps that are confidently wrong.

Clouds are the most obvious artifact. An optical satellite cannot see through clouds. The cloud pixel records the top of the cloud, not the ground beneath it. In a true-color composite, clouds look white and fluffy, exactly as you would expect. In false-color composites, they can take on teal or cyan hues, depending on band placement. The real trap is the cloud shadow, the artifact that fooled Priya's map. A shadow falls on the ground, not on the cloud layer, so the satellite records the actual ground surface under the shadow, but that surface receives far less sunlight than the surrounding scene. Every reflectance value in the shadow is depressed across all bands. Automated classifiers that learn from sunlit examples will often misread shadow regions as a different land cover type: forest where there is actually bare soil, water where there is actually farmland, stressed crops where there is actually healthy corn.

Diagram 4.3: The Four Artifacts Checklist

Sensor stripes are a different kind of problem. They arise from calibration drift in the detector array inside the sensor. Imagine the sensor as a row of thousands of tiny light-sensitive elements, each one responsible for recording one pixel column in the image. If even one element drifts out of calibration, it will consistently record slightly brighter or slightly darker than its neighbors, producing a visible stripe running the entire length of the image. The most famous example was the Landsat 7 scan line corrector failure in 2003, which introduced diagonal data gaps in every Landsat 7 image thereafter. Analysts working with that data learned to flag the affected pixels and fill them using neighboring imagery. Stripe artifacts are easy to recognize because they are perfectly straight and run in the along-track or across-track direction of the satellite's orbit, not along any natural boundary on the ground.

Mosaic seams appear when an analyst or a data provider combines images from multiple acquisition dates into a seamless-looking composite. Building a cloud-free view of a large area often requires stitching together dozens or hundreds of individual scene acquisitions taken over weeks or months. Each acquisition had a slightly different sun angle, atmospheric conditions, and sensor calibration. Even after correction, the seams where two scenes join can appear as a visible line of color discontinuity. You might see a forest that is bright olive on the left half of the image and dark forest green on the right, with a hard edge between them that does not follow any natural feature. The forest did not change at that line. The data sources did.

Saturated pixels occur when the reflected or emitted energy reaching the sensor exceeds the sensor's maximum recordable value. This is the satellite equivalent of overexposing a photograph, where the bright areas "blow out" to pure white. Bright desert sand on a cloudless day, fresh snow at high solar elevation, or the surface of an active lava flow can all cause certain sensor channels to saturate. A saturated pixel reports the sensor's maximum digital number, not the actual reflectance of the surface, so any analysis that uses that pixel is reading noise rather than signal. Good data providers flag saturated pixels in the quality mask that accompanies each image product. Responsible analysts mask them out before running any calculations.

Recognizing these four artifacts is a form of visual literacy that separates an informed reader from a naive one. The next time you see a satellite image in a newspaper and notice a perfectly straight diagonal stripe. This hard rectangular boundary aligns with nothing on the ground or with a region of pure white surrounded by reasonably colored areas; you will know what you are seeing. You are reading the picture, not just looking at it.

Atmospheric Correction in Plain English

Imagine trying to photograph a friend through a slightly fogged window. The glass blurs the colors, softens the contrast, and adds a faint milky veil to everything. You could take the photo anyway, or you could clean the window first. Atmospheric correction is the act of cleaning the window after the fact, in software, using physics to work backward from what the satellite recorded to what the ground surface actually looked like.

The atmosphere between the satellite and the ground is not empty. It contains gas molecules, water vapor, aerosols (tiny suspended particles), and sometimes thin cirrus cloud layers. These components interact with sunlight in two important ways. First, they absorb certain wavelengths, removing energy before it reaches the ground or before the reflected energy reaches the sensor. Second, they scatter light in all directions, which adds a diffuse veil of brightness to every pixel. The net effect is that a raw satellite image records a mixture of surface reflectance and atmospheric signal rather than pure surface reflectance.

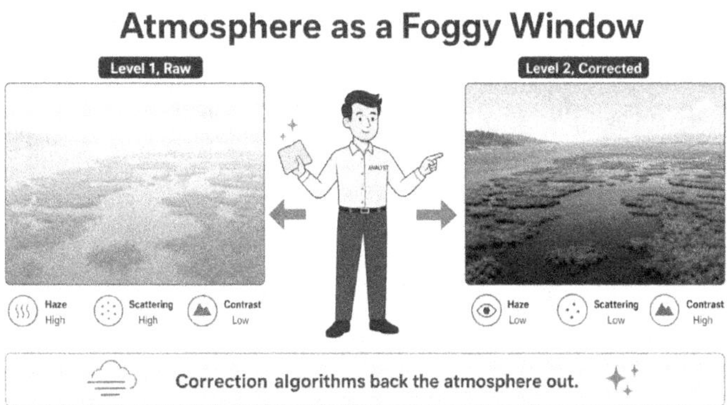

Diagram 4.4: Atmosphere as a Foggy Window

This is where the terms Level 1 and Level 2 become useful. A Level 1 data product, also called a top-of-atmosphere (TOA) product, contains the raw digital numbers recorded by the sensor after correcting for sensor geometry (ensuring each pixel maps to the correct location on the ground) but before removing the atmospheric contribution. The values in a Level 1 image represent what the sensor observed, including atmospheric effects. A Level 2 data product has been processed using an atmospheric correction algorithm to estimate and remove atmospheric effects, yielding values that represent the reflectance of the ground surface. Level 2 data is also called a surface reflectance product.

Why does the distinction matter? If you want to compare images of the same field taken in different months or different years, Level 1 data will mislead you. A hazy summer day will produce lower measured reflectances than a crystal-clear winter day even if the surface is identical, simply because more scattering occurred. Level 2 data removes most of that seasonal and atmospheric variability, making genuine temporal comparisons much more reliable. Priya's map showed a Level 1C product. Sentinel-2's Level 1C means geometry-corrected but not atmospherically corrected. For a vegetation index computation meant to track crop health, that was the wrong starting point.

Atmospheric correction algorithms work by modeling the atmosphere using additional inputs: the satellite's viewing angle, the solar elevation angle at the time of capture, estimated aerosol optical depth from ancillary sensors, and water vapor content derived from the image itself or from weather model outputs. The algorithm runs a physical model forward to predict what the sensor should have seen given those atmospheric conditions, then inverts the model to remove the atmospheric contribution from the recorded values. Different algorithms handle these steps differently, and the output is never perfect, but a good Level 2 product is almost always more useful for scientific analysis than the raw Level 1 counterpart.

One practical point for anyone consuming satellite-derived information: always ask which level you are looking at. If you are reading a report from a vendor, check whether they state the processing level of the underlying data. A product billed as a vegetation health map, computed from Level 1 data without any atmospheric correction, should carry a warning label. It can still be useful for a qualitative look at a scene, but the numeric values are not reliable enough for careful comparison across time or space.

Pixel Classification and Supervised Learning

Look at a satellite image of a city, and you will see residential streets, parks, industrial areas, rivers, parking lots, and airports all mixed. Every one of those land cover types has a characteristic spectral signature, a pattern of reflectances across multiple bands that distinguishes it from others. Forest reflects strongly in the near-infrared. Water absorbs nearly all incoming light across all bands and appears dark. Concrete reflects fairly uniformly across visible wavelengths. Bare soil has a rising reflectance profile from blue through shortwave infrared. The goal of pixel classification is to teach a computer to recognize patterns and label every pixel in an image with a land-cover category.

Supervised learning in this context works exactly the way its name implies. An analyst acts as the teacher, and the algorithm acts as the student. The teacher creates a set of training samples: a few hundred pixels that the analyst personally labels as water, a few hundred labeled as forest, a few hundred labeled as urban, and so on for however many categories the map needs. These labeled pixels form a training dataset. The analyst knows for certain what those pixels are because they have cross-checked them against aerial photographs, field observations, or other verified reference data. The training dataset is the foundation of the whole enterprise, and its quality determines the quality of everything that follows.

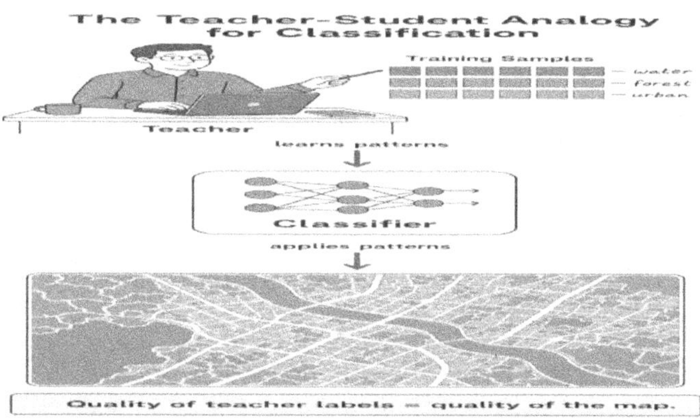

Diagram 4.5: The Teacher-Student Analogy for Classification

The student, in this case a classification algorithm, reads those training pixels and looks for the spectral pattern that distinguishes each category from the others. Different algorithms model that pattern differently. Some draw straight lines through spectral space to separate classes. Others build more complex boundaries based on probability. What they share is the core idea: given a labeled set of examples, learn the statistical characteristics of each category well enough to assign a label to a new, unseen pixel.

Once trained, the algorithm processes every pixel in the image, compares its spectral values with those learned from the training data, and assigns the most probable land-cover label. The result is a classification map: every pixel has a color code representing its assigned category. A well-trained classifier applied to a clear Sentinel-2 image of a temperate region can reliably distinguish open water, wetlands, forests, cropland, and urban areas with overall accuracy well above 85 percent. A poorly trained classifier applied to the same image might mistake parking lots for bare soil or sparse shrubland for urban gardens.

The word "supervised" distinguishes this approach from unsupervised classification, which asks the algorithm to find natural groupings in the data without any analyst labeling. In unsupervised classification, the algorithm clusters pixels by spectral similarity on its own. The clusters may or may not align with meaningful real-world categories. The analyst's job afterward is to interpret what each cluster represents, which is the reverse of the supervised workflow and often more ambiguous. Unsupervised methods are useful for exploratory analysis and for mapping areas where no training data exists, but they produce less reliable maps for operational applications.

Spotting a sloppy classification map is easier than you might think. The most common tell is "salt and pepper" noise: isolated single pixels of one class embedded in a solid field of another, with no spatial logic. A single pixel labeled "water" sitting in the middle of a desert, surrounded by pixels labeled "bare soil" on all sides, is almost certainly a classification error. Good mapmakers apply spatial smoothing filters after classification to eliminate these isolated outliers. Another tell is class boundaries that do not align with real topographic or land-cover edges. If the forest-to-cropland boundary in a map runs perfectly straight across a rolling landscape with no obvious field edge or road, the classifier may have learned an artifact rather than a real feature.

The term "supervised learning" here also serves as your introduction to a broader idea. The same concept of providing labeled training examples and letting an algorithm learn from them underpins most modern machine learning, from spam filters to medical diagnosis to the large language models that process text. In satellite image classification, the training labels come from an analyst with knowledge of the ground. In other applications, they come from human reviewers, historical records, or automated systems. The fundamental logic is the same: show the algorithm many labeled examples, let it extract patterns, then apply those patterns to new cases. Knowing that, the leap from "classifying pixels in a corn belt image" to "understanding how any AI classifies anything" is not as large as it first appears.

Object Detection: Ships, Planes, Buildings

Pixel classification labels each pixel in an image with a land-cover category. Object detection does something different. It looks for specific things, objects with recognizable shapes and spatial extents, and draws a box around each one. The output of a classifier is a wall-to-wall map of categories. The output of an object detector is a list of objects: fifty-two vehicles in this parking lot, twelve ships in this harbor, four aircraft on that runway.

The distinction matters because many useful questions in geospatial intelligence are object-level rather than area-level. A shipping analyst does not want to know what percentage of a port is covered by metal; they want to know how many ships are docked today compared to last Tuesday. A city planner does not need to classify every square meter as "building" or "road"; they need a count and the footprints of buildings to model population density. An emergency responder does not need a land-cover map of a disaster zone; they need to know how many structures have collapsed.

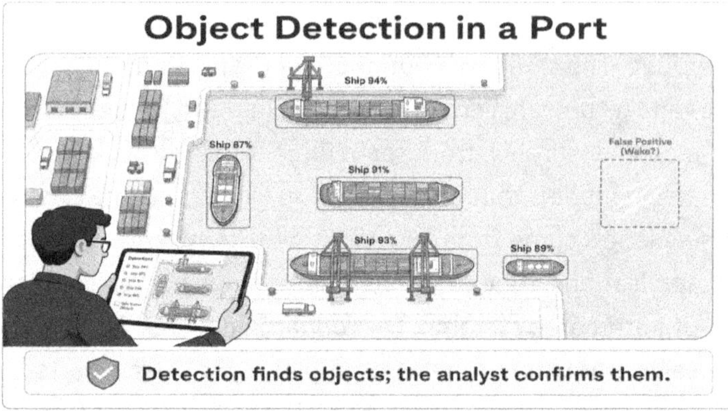

Diagram 4.6: Object Detection in a Port

Object detection in satellite imagery typically works by training a model on labeled examples of the object of interest. An analyst marks hundreds or thousands of ships in training images, each one outlined with a bounding box. The model learns the visual characteristics of a ship: the elongated hull shape, the bow wave at the front, the superstructure pattern, the relationship between length and width. Given a new image, the model slides a detection window across every location and scale, asks, "Does this look like the object I learned about?" and reports every location where the answer exceeds a confidence threshold.

Two numbers define the quality of an object detector. The first is precision, which answers the question: of all the detections the model reported, what fraction were correct? A precise detector rarely raises false alarms. The second is recall, also called sensitivity, which answers the question: of all the real objects in the image, what fraction did the model find? A high-recall detector misses very few objects. In practice, precision and recall trade off against each other. Lowering the confidence threshold catches more real objects (higher recall) but also catches more false positives (lower precision). Raising the threshold cuts false positives but lets more real objects slip through.

A third metric worth knowing is Intersection over Union, or IoU. When a model draws a bounding box around a ship, that box rarely perfectly matches the hand-drawn ground truth box used during evaluation. IoU measures how much the predicted and reference boxes overlap, divided by the area they cover together. An IoU of 1.0 means perfect overlap. An IoU of 0.5 means the predicted box covers exactly half of the combined area of the two boxes. Most object detection benchmarks report performance at a specific IoU threshold, such as "mAP at IoU 0.5," meaning the model is credited with a correct detection only if its box overlaps the reference box by at least 50 percent. Knowing that the IoU threshold exists helps you read the fine print of any vendor claim about detection accuracy.

Object detection has genuine limits that enthusiastic marketing materials often gloss over. Resolution matters enormously. A ship that spans dozens of pixels in a 50-centimeter-resolution image becomes a single blurry smudge in a 10-meter-resolution image. Aircraft on an airfield need at least 1-meter resolution to be reliably identified by model type rather than "there is a plane-shaped object here." Small boats, individual vehicles, and building entrances typically require sub-meter resolution. When a vendor claims their algorithm can count fishing boats across an entire ocean using freely available imagery, ask about the resolution and the documented false positive rate before believing the headline number.

Real-world applications of object detection include counting ships in contested maritime zones to estimate shipping activity, monitoring construction activity in restricted areas over time, counting vehicles in hospital parking lots to estimate patient volumes, and locating survivor signatures in search-and-rescue operations following earthquakes. Each application has its own resolution requirements, tolerance for false positives, and need for human review before acting on the model's output. The detector provides a first-pass screening. The analyst, or the client, provides the judgment.

Ground Truth and Why You Cannot Skip It

Every satellite analysis makes a claim about the real world. A classification map claims that this pixel is forest and that pixel is cropland. An object detector claims that this elongated shape is a ship. A vegetation index map claims that this field is stressed. Ground truth is the process of checking those claims against direct observation of the real world. It is the mechanism by which responsible practitioners measure how much to trust a satellite-derived product.

The term "ground truth" is borrowed from survey and cartography, where it refers to an observation made directly on the ground to confirm what a map or model predicts. In satellite analysis, it has the same meaning. An analyst travels to a sample of classified pixels, observes what is actually there, and records whether the classification was correct. If the map said "forest" and the ground observer found "forest," that is a correct match. If the map labeled the area as forest and the ground observer found a recently harvested clearing, that is a commission error: the classifier assigned a category to a location where it does not apply.

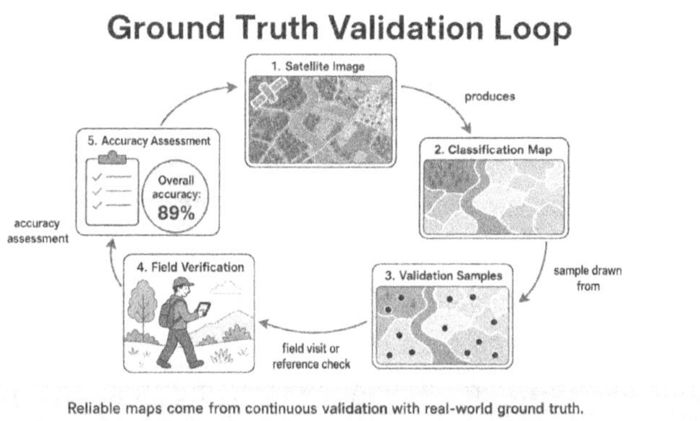

Diagram 4.7: Ground Truth Validation Loop

Accuracy assessment follows a statistical logic. You cannot visit every pixel, so you draw a random sample of classification pixels and verify them from that sample, you can estimate overall accuracy (what percentage of all classified pixels are correct), producer's accuracy (for a given reference category, what fraction did the classifier correctly identify), and user's accuracy (for a given map category, what fraction are actually that category on the ground). These three numbers tell different stories. A map might show 95 percent overall accuracy but have only 60 percent user accuracy for wetlands, meaning four out of ten pixels labeled "wetland" are actually something else. For applications where wetland mapping matters, that is not good enough.

A responsible geospatial product will always include an accuracy assessment. If you see a classification map, a land cover dataset, or a change detection product without a documented accuracy assessment, treat it with healthy skepticism. The absence of validation does not necessarily mean the product is bad; it simply means you have no information to calibrate your trust. Asking "what was the validation methodology?" is one of the sharpest questions a buyer or a journalist can ask of any geospatial vendor.

Priya knew this intuitively from years of working with corn and soybean growers. She would not deliver a vegetation stress map to a client without walking a sample of the flagged fields herself, or at least asking the grower to confirm on the ground. Not because she doubted her satellite pipeline, but because field conditions can change between image acquisition and client delivery, and because a single unresolved artifact, such as a cloud shadow, could cost a grower an unnecessary intervention. Ground truth was her quality check and her professional insurance.

For applications with higher stakes, the validation standard is correspondingly stricter. A deforestation monitoring service that feeds national carbon accounting programs must demonstrate validation methodology that meets international reporting standards. A battlefield damage assessment that informs a military response needs not just accurate statistics but a clear chain of evidence showing which pixels were verified, how, and when. In both cases, the principle is the same: the satellite is a powerful first observer, but the real world is the final authority.

How to Read a Satellite Story in the News

Satellite imagery appears in the news more than ever. Coverage of wars, floods, wildfires, crop failures, prison camps, factory expansions, and climate change increasingly uses satellite-derived evidence as primary sourcing. This is a healthy development for public information. It is also a space where misleading claims, unsupported interpretations, and outright fabrications can circulate because most readers lack a framework for evaluating what they see.

The checklist that follows is designed to give you that framework. It is not a checklist for experts. It is five questions that a curious and careful reader can ask in under two minutes, looking at any published satellite image or satellite-derived claim.

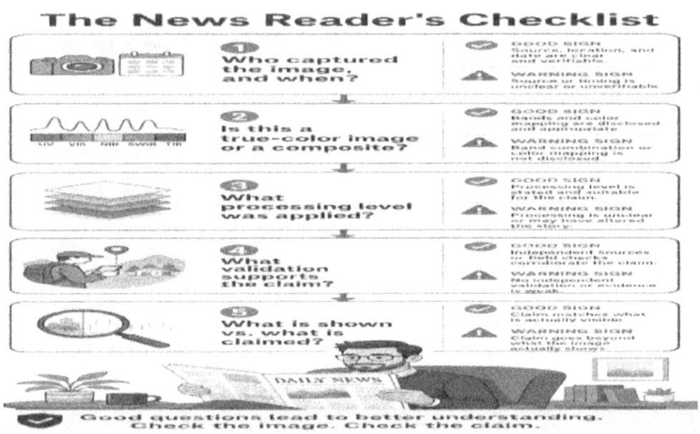

Diagram 4.8: The News Reader's Checklist

The first question is: who collected this image and when was it captured? Source and date are the most basic metadata for any image, and their absence is a red flag. A reputable organization, whether a commercial satellite operator, a government agency, a university, or an established investigative outlet like the New York Times Visual Investigations team or Planet's open data program, will always document the satellite name, the capture date, and ideally the exact time. An image attributed solely to "satellite data," with no further specification, should prompt you to look harder before accepting the claim it illustrates.

The second question is: is this a true-color image or a processed composite? A true-color image in standard red-green-blue bands will look roughly like what you expect from a photograph. If the vegetation is bright red, you are almost certainly looking at a false-color or false-color infrared composite. If the ocean is black and the desert is vivid orange, some other band combination was chosen. None of that is dishonest. But the image caption should tell you what you are looking at. If the caption claims the image "clearly shows" something in a color that does not match natural appearance, ask whether the color scheme itself was chosen to make that thing visible.

The third question is: what processing level was applied, and is the claim compatible with that level? Many published satellite images are Level 1 or equivalent, meaning they retain atmospheric effects and may not be suitable for precise quantitative claims. A statement like "our analysis shows vegetation density decreased by 12 percent over this region" requires Level 2 surface reflectance data and a rigorous index calculation, not a Level 1 visual composite. If the published claim is quantitative, the underlying data should be at least Level 2, and the methodology should be documented.

The fourth question is: what ground validation supports the interpretation? A satellite image might show what appears to be construction activity at a suspected facility. Has anyone verified, from open-source records, commercial imagery from a second satellite, or on-the-ground reporting, that construction actually occurred? Satellite evidence is strongest when it corroborates other evidence and weakest when it stands alone. Reputable investigative teams using satellite imagery, such as Bellingcat, the New York Times Visual Investigations unit, or Human Rights Watch's satellite analysis program, routinely document their validation methods. When that documentation is present, the claim is on much firmer ground.

The fifth question is: what exactly is being claimed versus what the image actually shows? A common pattern in misleading satellite coverage is the gap between what the image physically shows and the interpretation placed on it. An image of a construction crane at a port proves only that there is a crane, not that a weapons system is being assembled. An image of cleared land in a rainforest shows that land has been cleared, not the specific cause or actor behind the clearing. The image is evidence toward a claim. The claim requires additional logic. When a headline says "satellite images prove X," find the actual image and ask what it proves on its own, without the editorial frame.

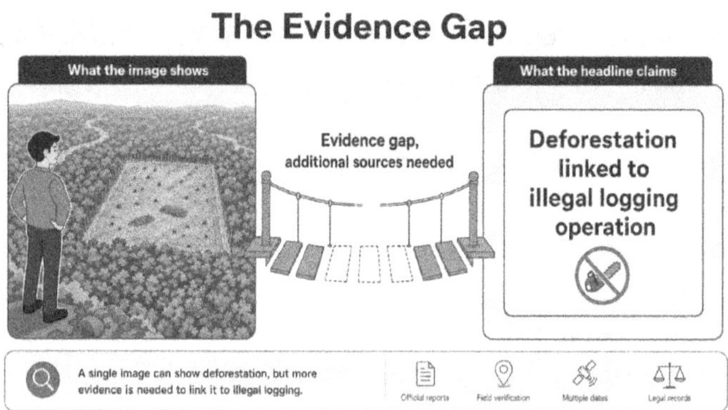

Diagram 4.9: The Evidence Gap

You do not need to be a remote sensing scientist to apply this checklist. You need to be the kind of reader who pauses before accepting a dramatic claim and asks the same basic questions you would ask of any news source: Who said this? When? Based on what? Checked how? Claimed to show what?

Takeaway

Reading a satellite image well is not a mystical skill. It is a set of habits. Ask what the colors represent. Check the processing level in the metadata. Look for the four telltale artifacts: cloud shadows hiding behind alarmed red pixels, sensor stripes running ruler-straight across a mosaic, seam lines that cut across natural features, saturated whites where the sensor overloaded. Ask who trained the classifier and what their training data looked like. Ask where the ground truth is.

Those habits fit together into something larger than individual picture-reading. They are a posture toward satellite-derived information, one that is neither dismissive nor credulous. Dismissive readers ignore satellite evidence in public debates where it is genuine and useful. Credulous readers accept every colorful composite as unmediated reality. The reader who understands false-color composites, artifact recognition, and accuracy assessments sits in the productive middle ground. That reader can evaluate a claim, ask the right follow-up, and participate in the conversation with genuine confidence.

Daniela Reyes did not develop her three-question diagnostic overnight. She built it from experience, from maps that were wrong before the questions were asked, from crops misdiagnosed, fields mismanaged, and reports embarrassingly retracted. The questions are now reflexes. They will become yours too. The satellite is honest. Reading it is a skill.

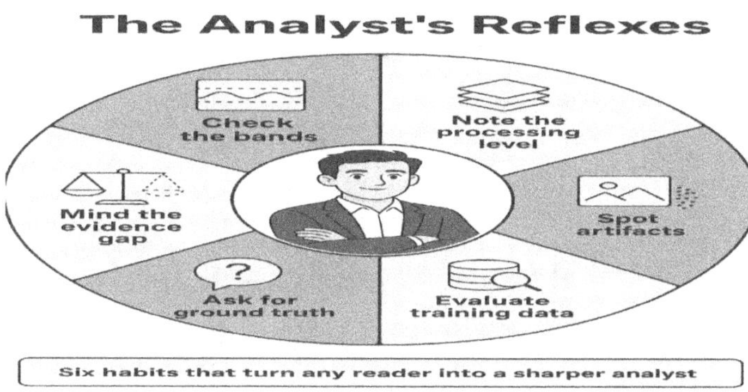

Diagram 4.10: The Analyst's Reflexes

The physics of light, sensors, and orbits gives geospatial intelligence its raw material. The skills in this chapter are what turn that raw material into trustworthy knowledge. Image interpretation is where the satellite conversation leaves the lab and enters the world. It is where Priya's map either correctly guides a grower's decision or sends them chasing a cloud shadow. It is where a newsroom's satellite evidence either informs the public or misleads them. It is, in the end, where the technology earns its value or loses it.

The next chapter looks at what happens when the analysis scales beyond a single image, across thousands of images, to monitor the same place over months and years. Time series analysis opens a different kind of reading, one where change itself is the signal.

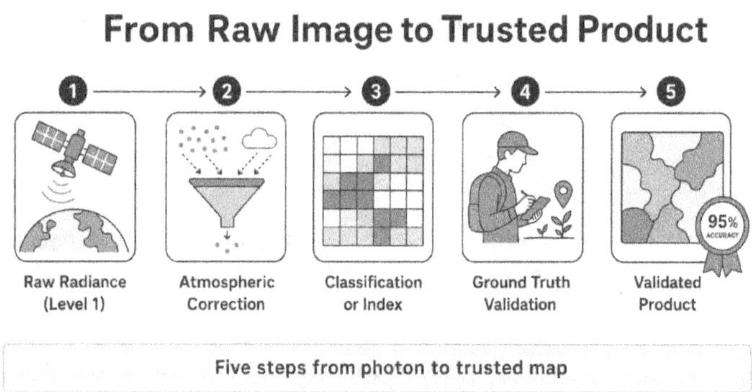

Diagram 4.11: From Raw Image to Trusted Product

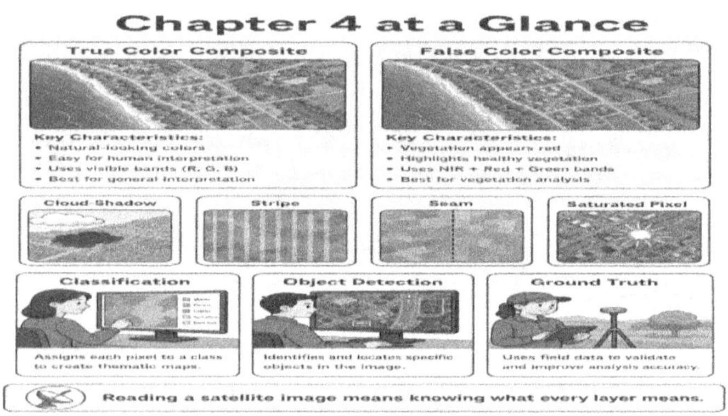

Diagram 4.12: Chapter 4 at a Glance

5 Watching the World Change

Scenario

Dr. Amara Wells arrives ten minutes early and spends most of that time adjusting the projector brightness. The room is a county boardroom overlooking the Gulf, and the twelve people filing in are coastal managers, port directors, and one very quiet state senator. They have been arguing about beach erosion for three years. Amara has a single animation file on her laptop.

She connects the cable, dims the overhead lights, and opens the file without commentary. The screen shows an aerial view of a barrier island, and then she presses play. Forty years compress into forty seconds. The western tip of the island, the part that juts toward the shipping channel, begins to move. Not dramatically, not like a movie explosion. Just a slow, unmistakable retreat. The pale sand that used to extend into the frame disappears, year by year, pixel by pixel, until the 2023 frame shows open water where a parking lot and a pavilion once stood.

Nobody says anything for several seconds. Then one of the port directors exhales slowly and says, "Can we run that again?" Amara does. She runs it three times. By the fourth pass, the room is leaning forward in their chairs. This is what change detection does. It does not argue. It does not editorialize. It lets time speak, and time is a very persuasive witness.

Why It Matters

Every chapter up to this point has been about what a single image shows: the bands of light it captures, the features it can resolve, the corrections applied to make it accurate. That single-image thinking is powerful, but it is also limited. A single image is a photograph. It shows you the world at one moment and nothing more.

Change detection is the moment when remote sensing stops being photography and becomes more like cinema. The instant you have two images of the same place taken at different times, you can ask a question that a photograph can never answer: what changed? And the instant you have hundreds of images spanning years or decades, you can ask something even richer: is this place trending in a direction? Is it warming, greening, shrinking, flooding more often?

That shift from snapshot to sequence is what this chapter is about. It is the conceptual leap that turns a pile of satellite images into a monitoring system. And once you understand how it works, you will start to see the world differently. You will read news stories about deforestation, glacial retreat, or urban expansion and understand exactly what kind of analysis sits behind those claims. You will know how to ask for the data, what to trust, and what traps to watch for.

Before-and-After and Why It Works

The most intuitive form of change detection is also the most ancient. You look at one picture, then another taken later, and you note the differences. Archaeologists have done this for decades with aerial photographs. Insurance adjusters do it with before-and-after storm photos. Your brain is actually exceptionally good at this task. Given two nearly identical images, the human visual system notices differences almost instantly, the same way it spots a moved piece of furniture the moment you walk into a room.

Satellite imagery formalizes that instinct into a quantitative operation. The basic idea is straightforward: take the pixel values from image A, subtract the corresponding pixel values from image B, and examine the result. Where nothing changed, the subtracted values are near zero. Where something changed significantly, the values are large, either positive or negative depending on whether the pixel got brighter or darker. The resulting "difference image" is a map of change, coded by magnitude.

Diagram 5.1: Before, After, and the Difference Image

What makes this technique so persuasive is its visual immediacy. When a coastal management board sees two images of the same beach, taken 18 years apart, and the waterline has moved 50 meters inland, they do not need a statistics lecture. The evidence is spatial. It lives in the frame. You can pinpoint the exact spot where the change occurred, measure it, and watch it unfold year by year.

The technique has fueled some of the most consequential environmental journalism of the past two decades. Deforestation monitoring in the Amazon basin depends on exactly this approach: analysts compare this year's dry-season image to last year's dry-season image, look for newly bare-soil pixels where forest used to be, and count the hectares. The same logic applies to post-hurricane damage assessments. Before a storm, FEMA-coordinated analysts record the baseline structure of a coastal city. After the storm, they overlay the new image on the old one and generate a damage footprint map within hours, guiding rescue teams on where to go first.

The "before-and-after" framing is so effective that it has become the default communication tool for almost every satellite-imagery story that reaches general audiences. News organizations routinely publish interactive sliders that let readers drag a dividing line between the two dates, revealing change in real time. The spatial specificity of the comparison does something that charts and tables cannot: it grounds the change in a place that people recognize as real.

That said, before-and-after differencing is the bluntest instrument in the analyst's toolkit. It captures any change between the two dates, whether meaningful or not. A field of corn photographed in May and September will look dramatically different simply because the crop grew and then was harvested. That is not an environmental disaster; it is agriculture. A desert photographed before and after a rainstorm looks like it greened up overnight. That is not vegetation recovery; it is wet soil. The technique works best when you have strong domain context for what kinds of change to expect and can choose your image dates carefully to minimize seasonal confounding. That context-setting is part of the craft.

The workhorse status of before-and-after differencing in disaster response comes from one practical virtue: speed. When a city floods, analysts do not have weeks to run elaborate trend models. They need a damage footprint in hours. Two images and a subtraction get them there. More sophisticated methods require more time and more data. So the field uses the right tool for the urgency of the moment, and for rapid operational response, the simple before-and-after difference is often exactly what is needed.

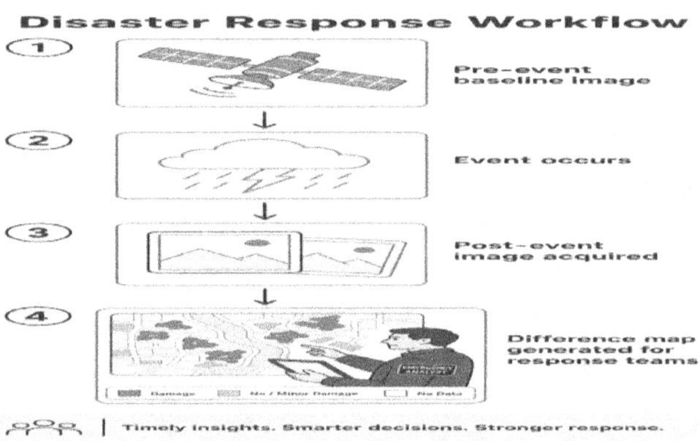

Diagram 5.2: Disaster Response Workflow

The Planet's Heartbeat

If before-and-after differencing is a snapshot comparison, time series analysis is something richer. A time series is exactly what it sounds like: the same measurement taken repeatedly at regular intervals over a long stretch of time. Your bathroom scale is running a time series if you weigh yourself every morning and record the result. A temperature sensor at a weather station is running one. The key property is regularity: same place, same measurement, over and over.

Satellites are exceptionally well-suited to generating time series. Landsat has been imaging the entire Earth's land surface every sixteen days since 1972. That is more than fifty years of coverage at 30-meter resolution, a scientific asset of extraordinary depth. Sentinel-2, the European counterpart, images every part of Earth's land surface every five days at resolutions down to 10 meters. Other satellites add radar, thermal, and higher-resolution optical layers. The result is a planet wrapped in sensors, generating regular repeated measurements of its own surface condition.

When you stack all those measurements for a single pixel, say one 30-meter-by-30-meter patch of forest in central Africa, you get a time trace. The x-axis is time. The y-axis is whatever measurement you chose: reflectance in the near-infrared band, a vegetation index, surface temperature. The line that emerges is not random. It has structure. It rises and falls with the seasons. It dips during droughts. It recovers after disturbance. It tells a story about that patch of ground that no single image could ever tell.

This is the electrocardiogram (EKG) analogy, and it is not just a rhetorical flourish. An EKG measures the heart's electrical activity over time and records it as a waveform. A cardiologist reading that waveform is not looking at any single heartbeat in isolation. They are reading the pattern: the rhythm, the shape of each beat, the intervals between beats, the presence of irregularities. The waveform across time is diagnostic in a way that a single measurement never could be.

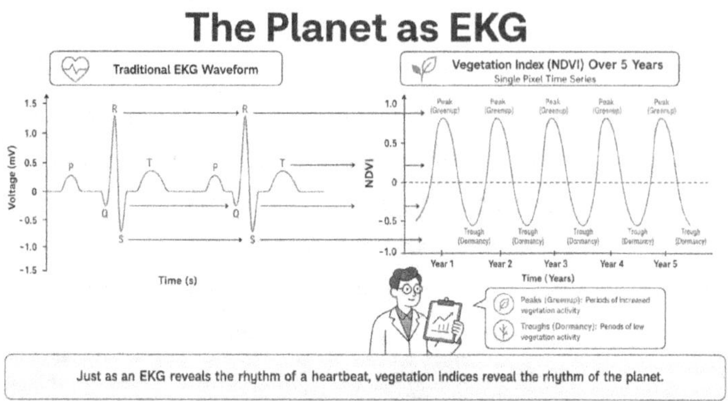

Diagram 5.3: The Planet as EKG

The planet's time trace has a heartbeat, too. Every ecosystem pulses through the year. Boreal forests green up in late spring and shed leaves in autumn. Grasslands in seasonal climates flush during the wet season and dry to straw by the end of the dry season. Croplands show sharp transitions as fields are planted, grow, and are harvested in patterns specific to the crop and the farm management practices. Snow accumulates through winter and melts in spring, feeding rivers at predictable intervals. Each of these rhythms leaves a clear signature in a time series.

The medical-imaging analogy is even richer when applied to long-term change. A single MRI scan shows you the state of a tissue at one moment. It might look normal. But a sequence of MRI scans taken over months can reveal whether a suspicious region is growing or stable, something no single image could tell you. Satellite time series do the same thing for the planet. A single image of the Greenland ice sheet shows you its size today. A forty-year time series of the same area shows you whether it is shrinking, how fast, and whether the rate of change is accelerating. The sequence is diagnostic. The snapshot is not.

This is why the phrase "Earth observation" is slightly misleading when taken at face value. It suggests passive watching, taking a picture and filing it. What modern Earth observation systems actually do is closer to continuous monitoring, more like a hospital's vital-signs telemetry than a photograph album. The value is not in any individual image. The value is in the unbroken sequence, the time series that never stops running.

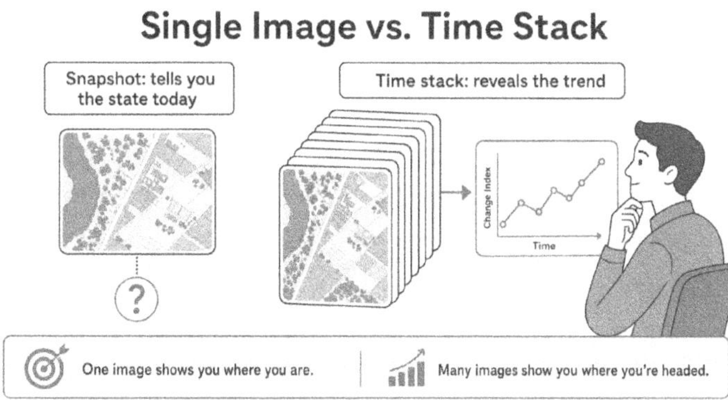

Diagram 5.4: Single Image vs. Time Stack

Landsat's fifty-year archive has enabled some of the most important environmental research in history. Studies of Amazon deforestation, Arctic warming, lake shrinkage, urban expansion, coral reef bleaching, and agricultural intensification all rely on it. The archive is free and public, a genuine gift from decades of government investment. The fact that you can download a complete time series for any pixel on Earth's land surface, for free, at 30-meter resolution, stretching back more than half a century, is genuinely remarkable. It is one of the great scientific infrastructure investments of the twentieth century, and its dividends compound every year as more time is added to the series.

Fire Scars, Floods, and Purposeful Indices

Generic differencing (band value from time A minus band value from time B) is a blunt instrument. It works when the change is large and spectrally obvious, such as a forest becoming a parking lot. But many kinds of change are subtler, and they require dedicated indices designed to amplify the specific spectral signal that accompanies the change you are looking for.

Chapter 4 introduced the concept of spectral indices through NDVI (Normalized Difference Vegetation Index), which exploits the contrast between visible red reflectance (low for healthy vegetation) and near-infrared reflectance (high for healthy vegetation). The same design logic applies to change detection. You build an index that is mathematically tuned to respond strongly to the kind of change you care about and weakly to the kinds of variation that are just noise.

Fire is one of the events that satellite-based change detection is best at documenting. When vegetation burns, the spectral signature of the land surface changes dramatically. Green, photosynthetically active plants reflect strongly in the near-infrared and absorb in the red. After a fire, the charred ground and dead vegetation look completely different. The near-infrared reflectance drops sharply. The shortwave-infrared reflectance changes in a characteristic way. The NBR, or Normalized Burn Ratio, is the index designed to capture this.

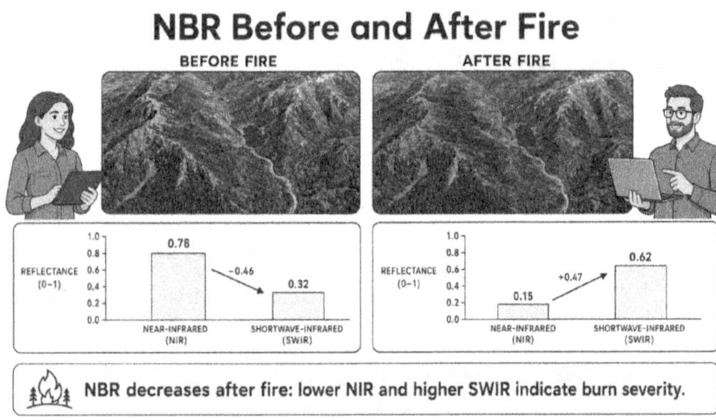

Diagram 5.5: NBR Before and After Fire

NBR uses the near-infrared (NIR) and shortwave-infrared (SWIR) bands. The formula is (NIR minus SWIR) divided by (NIR plus SWIR), the same normalized-ratio structure as NDVI. Unburned healthy vegetation typically produces a high NBR value, often near 0.5 or higher. Burned areas produce a much lower NBR, sometimes below zero. The difference between the pre-fire NBR and the post-fire NBR, called dNBR (delta NBR), is the standard measure of burn severity. A small dNBR means the fire barely touched the area or burned at low intensity. A large dNBR means severe burning: high-temperature crown fires that killed mature trees and scorched the soil.

The US Forest Service and national park managers use dNBR to map post-fire severity across entire mountain ranges within days of a fire's containment. They use those maps to prioritize watershed rehabilitation, plan erosion-control work, and understand where vegetation recovery efforts are most needed. A single Landsat or Sentinel-2 overpass after the smoke clears can automatically produce a fire-severity map covering hundreds of thousands of acres at 30-meter resolution.

The dNDVI, or delta Normalized Difference Vegetation Index, serves a similar purpose but for vegetation loss more broadly. It is not specific to fire. A forest that is deforested, a cropland hit by a hailstorm, a grassland overgrazed to bare dirt, or a wetland drained all show up as negative dNDVI events: the post-event NDVI is substantially lower than the pre-event NDVI. dNDVI is one of the workhorses of deforestation monitoring systems because it is simple, globally applicable, and responds to any vegetation removal regardless of the cause.

Floods present a different spectral challenge. Floodwater is not always apparent in optical imagery, especially if the flooding occurred under cloud cover (which is common, since floods often result from rain, and rain often comes with clouds). This is one reason radar-based change detection is so important for flood mapping. Radar signals behave very differently over dry land versus open water. Synthetic aperture radar (SAR) sensors can see through clouds and show the extent of flooding even while the storm is still overhead. When analysts create water-difference maps using before-and-after SAR imagery, they are essentially asking: which pixels that were dry before the event are now wet? The result is a flood-extent map that goes straight into emergency management decisions.

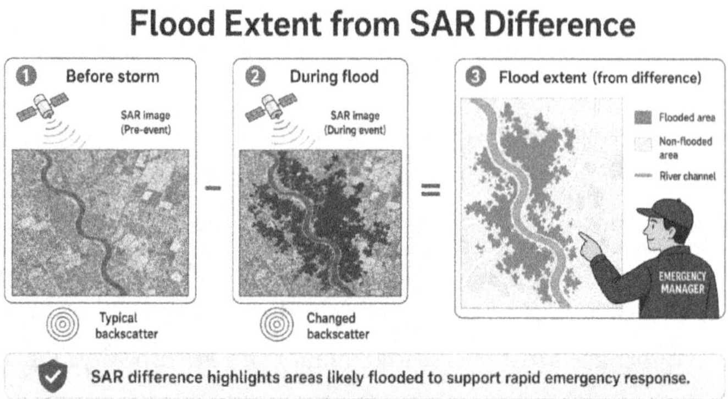

Diagram 5.6: Flood Extent from SAR Difference

The common thread across all purpose-built indices (NBR, dNBR, dNDVI, water difference) is spectral specificity. Each one is engineered to be sensitive to the exact spectral changes that accompany a particular event while being relatively insensitive to confounding signals. This is not unique to remote sensing. It is the same engineering logic that goes into a medical diagnostic test: design the test to respond to the presence of the condition you are looking for and to ignore everything else. The more precisely you can specify the spectral fingerprint of your target change, the more reliably your index will detect it and the fewer false alarms it generates.

Phenology: The Seasonal Heartbeat of Vegetation

Phenology is the study of the timing of biological events. When do buds break in spring? When does a forest reach its peak leaf-area and deepest green? When do crops flower? When does the chlorophyll drain out of autumn leaves and the canopy shift from green to orange? These events are not random. They follow seasonal patterns driven by temperature, day length, and rainfall, and they have been documented by naturalists for centuries. Henry David Thoreau recorded the flowering dates of plants around Walden Pond year after year in the 1850s, and those notebooks are now a precious dataset in climate-change research.

Satellite remote sensing allows phenology to be tracked at the scale of entire continents instead of a single pond. The tool is the NDVI time series. Because NDVI responds to vegetation's greenness and photosynthetic activity, a year's worth of NDVI measurements for any pixel traces the vegetation's life cycle with remarkable fidelity. NDVI rises steeply in spring as leaves emerge. It plateaus at a high summer maximum. It declines in autumn as leaves age and fall. It hits a winter minimum of bare soil or dormant grass. This annual curve is called the phenological profile, and it is as characteristic of an ecosystem as a fingerprint.

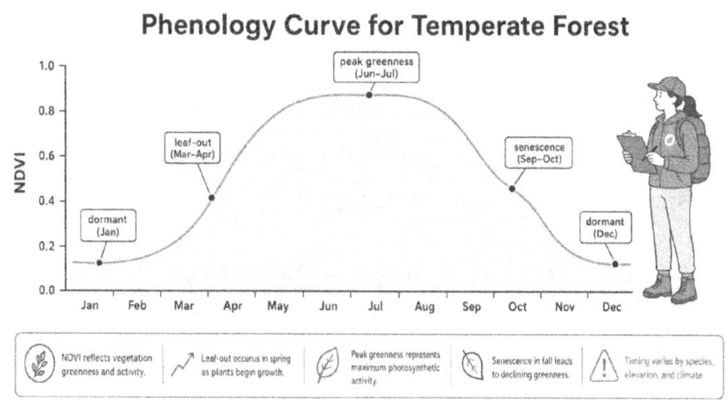

Diagram 5.7: Phenology Curve for Temperate Forest

What makes phenology so scientifically important is that it changes. Over the past several decades, the phenological profiles of many ecosystems have been shifting in consistent ways. Spring green-up is arriving earlier. Autumn senescence is arriving later. In many temperate regions, the growing season is measurably longer than it was thirty years ago. These shifts are consistent with a warming climate, and the satellite record is one of the most important sources of evidence for documenting them at global scale.

For agriculture, phenology tracking is both a diagnostic tool and a planning aid. Priya Bhatt, the precision agriculture lead at Pleiades Geospatial Services, spends much of every growing season comparing the current phenological state of her clients' fields against historical baselines. If a corn field's NDVI trajectory in July lags the historical median for that location by two weeks, that is a signal: either the planting was late, the seeds are underperforming, stress is occurring, or some combination of these. The phenological curve gives her a benchmark and a deviation to investigate.

Ecologists use the same logic to monitor the boundary between biomes, the places where the boreal forest gradually gives way to the Arctic tundra, for example, or where Mediterranean scrubland meets the edge of the desert. These boundaries, called ecotones, are among the most sensitive indicators of climate change. When a border that held its position for centuries begins moving northward or upslope, that movement shows up in the phenological data before almost any other measurement. The satellite time series is the instrument that can track a shifting treeline across a mountain range spanning thousands of kilometers, with updates every two weeks.

Crop calendars are another application. Different crops in different regions follow distinct phenological schedules: when they are planted, when they canopy close, when they flower, and when they are harvested. Analysts can use time series of NDVI and related indices to build crop-specific phenological templates, then use those templates to automatically classify which crop is growing in each field, year by year, across an entire country. This is how global crop-monitoring agencies produce early estimates of food production before the harvest even occurs. The signal is in the timing, and the timing is in the time series.

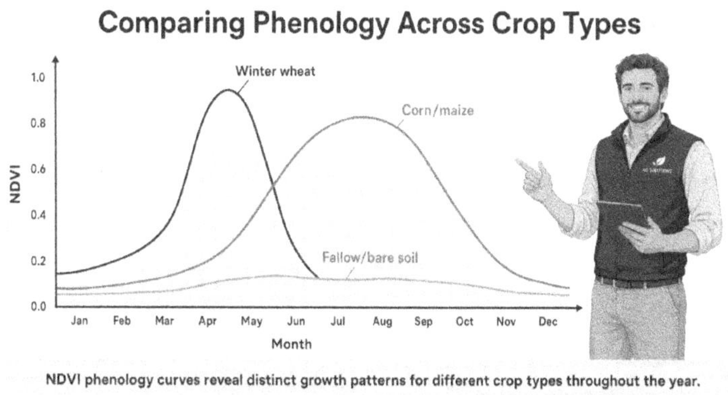

Diagram 5.8: Comparing Phenology Across Crop Types

The phrase "seasonal heartbeat" is deliberately evocative. Just as a cardiologist learns to read a patient's cardiac rhythm and notice any irregularity in the beat, an ecologist or agronomist trained to read phenological profiles learns to notice when the vegetation's annual rhythm is off. Early senescence can indicate drought stress. A delayed green-up can mean a late frost. A flattened summer peak can mean a pest outbreak or disease pressure. The deviation from the expected rhythm is the signal. The time series is the instrument that makes that rhythm visible.

Trend Detection Without Statistics Jargon

Seasonal cycles are real and expected. Summer is greener than winter. The crop grows and then is harvested. Snow falls and then melts. A good analyst has to distinguish between this cyclical variation and genuine sustained change. That is the central challenge of trend detection, and it is harder than it sounds.

Think of the thermostat in your house. Your home's temperature cycles up and down every day. It is warmer in the afternoon than at midnight. It is warmer in summer than in winter. If you looked at only one data point, you could not tell whether your house is trending toward a long-term average that is higher or lower than it used to be. But if you measure the temperature at the same time every day for a year, you can start to pull apart the cyclical variation from the underlying trend.

"Is the baseline temperature of my house drifting upward?" is a trend question. "Why is it hotter at 3 pm than at 3 am?" is a cycle question. They require different analytical approaches.

Satellite time series analysts face the same challenge. Every pixel on Earth has a seasonal cycle baked into its reflectance values. Separating that cycle from a real, sustained directional change requires a method that first models seasonality and then looks for departures from the expected seasonal pattern. Two algorithm families have become the standard tools for this: BFAST and CCDC.

BFAST stands for Breaks For Additive Seasonal and Trend. The name is descriptive: the algorithm takes a time series, fits a model that includes both the seasonal cycle and a long-term trend component, and then looks for "breaks," or sudden discontinuities in either component. A break in the trend might represent deforestation or land conversion. A break in the seasonal component might represent a change in vegetation type or a shift in the crop calendar. BFAST does not require you to tell it when the break occurred. It finds the break by systematically searching the time series and asking: at which point in time does adding a break most improve the model's fit to the data?

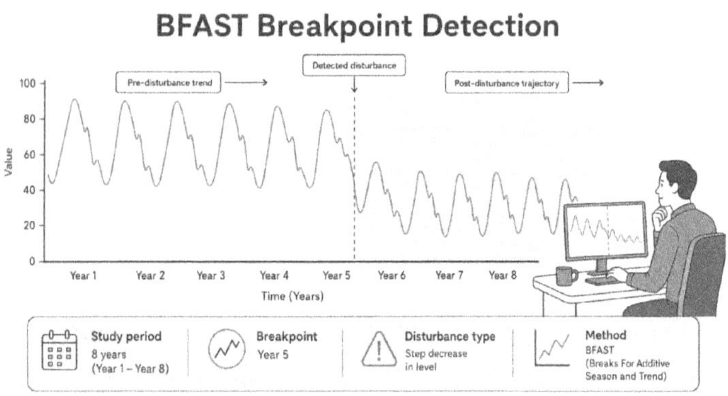

Diagram 5.9: BFAST Breakpoint Detection

CCDC stands for Continuous Change Detection and Classification. Where BFAST is commonly applied in batch mode to a pre-collected time series, CCDC is designed to run continuously, comparing each new incoming observation against a model built from the history of that pixel. Think of it as a persistent watch system: for every pixel, CCDC maintains a running model of the expected value, given the season, the time of day, and the longer-term trend. When new observations deviate from the model's predictions by more than a specified threshold for a specified number of consecutive observations, the algorithm flags a likely change.

The home-security analogy helps here. Imagine a motion-sensitive camera in your front yard. It is not alarmed by the regular comings and goings that it has learned to expect: the mail carrier at 10 am, your car leaving in the morning and returning in the evening. It has built a model of normal activity. It flags the unusual: movement at 3 am, movement from an unfamiliar direction. CCDC does the same thing for pixels. It learns the normal seasonal cycle and trend. It flags the anomalous. And because it runs in near real time on each new satellite pass, it can generate alerts for recent deforestation events or other land changes within days of their occurrence.

Neither BFAST nor CCDC requires the analyst to do the mathematical derivation by hand. Both are implemented in open-source libraries and platforms such as Google Earth Engine, where analysts can apply them at continental or global scales. The value is conceptual: understanding that change detection, at its most sophisticated, is not just subtracting two images. It models the expected behavior of every pixel over time. Then it asks which pixels are behaving in ways that the seasonal cycle, sensor calibration, or random noise cannot explain.

Other algorithm families fill specific niches. LandTrendr (Landsat-Based Detection of Trends in Disturbance and Recovery) is specifically designed for forest disturbance and recovery trajectories, tracking not just when a disturbance occurred but also how the forest has recovered in the years that followed. COLD (Continuous Monitoring of Land Disturbance) builds on CCDC concepts with additional robustness to cloud contamination. The specific algorithms matter less for most readers than the core idea: modern trend detection is about modeling time, not just comparing two dates.

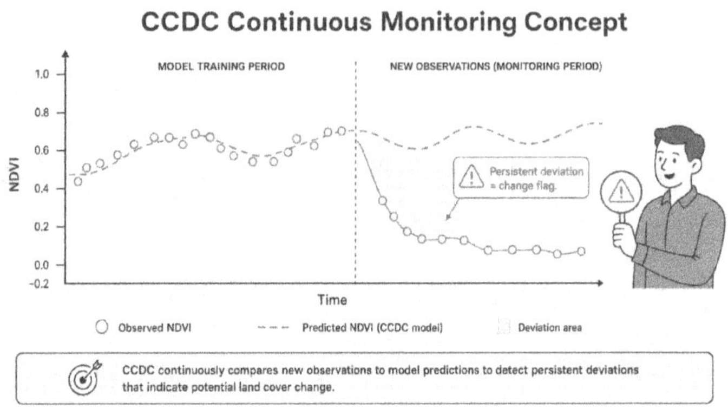

Diagram 5.10: CCDC Continuous Monitoring Concept

The Interpretive Traps of Change Detection

Change detection is one of the most powerful tools in remote sensing, and it is also one of the easiest to misuse. Three traps consistently catch beginners, and experienced analysts still fall into them occasionally under time pressure.

The first trap is confusing seasonal cycles with permanent change. This is the most common mistake and the easiest to make when you are working with only two dates. Imagine you compare a satellite image of farmland taken in November with one taken in May. In November, the fields are bare after harvest. In May, they are covered in dense green crops. The difference image looks dramatic: huge areas of the landscape appear to have been "cleared" or "greened up." But nothing has permanently changed. The fields are simply in different phases of the same annual cycle. This mistake sounds obvious when described this way, but in practice, analysts under deadline pressure working with unfamiliar landscapes make it regularly. The discipline is always to check the dates of both images against the known seasonal calendar for that ecosystem, and when possible to compare images taken at the same phenological phase in different years rather than two randomly dated images.

The second trap is sensor calibration drift. Satellites age. Their sensors drift slightly over time as the optical components are exposed to radiation in space. If you compare an image from a freshly launched satellite to one from the same satellite ten years later, you may see radiometric differences that have nothing to do with what happened on the ground. The ground might be completely unchanged, but the sensor is measuring slightly differently. Responsible time series analysis requires careful radiometric calibration and, in many cases, cross-calibration between different satellite missions to ensure that differences in the time series are real changes rather than sensor artifacts. The Landsat archive is radiometrically calibrated to very high standards, but even with excellent calibration, careful analysts still validate their change signals against independent measurements and known stable reference sites.

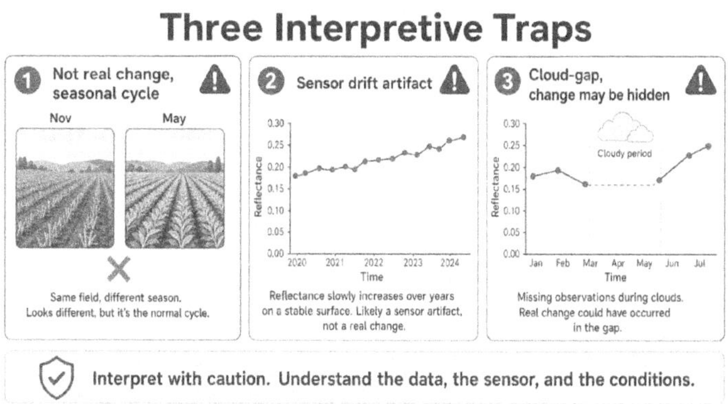

Diagram 5.11: Three Interpretive Traps

The third trap is cloud-gap artifacts. Optical sensors cannot see through clouds. In persistently cloudy regions, tropical rainforests, monsoon landscapes, and high-latitude winter zones especially, usable cloud-free imagery may be available only a few times per year. If a major change event (e.g., a deforestation pulse, a flood, or a fire) occurs during a cloudy period, the optical time series may show a jump from "before the event" to "after the event" with no imagery in between. The analyst sees the before-and-after correctly, but the timing is uncertain. Worse, if the analyst's time series includes an image acquired right after the cloudy gap and the change is incorrectly attributed to the wrong date, it can propagate into downstream analyses and alerts.

Radar-based change detection partially addresses this problem because it can see through clouds. But radar and optical sensors measure different things, and combining them requires additional calibration care. The honest answer is that cloud gaps are a permanent limitation of optical satellite time series and one that analysts must account for in every analysis. They do this partly through algorithm design (CCDC, for example, requires persistent deviations across multiple consecutive observations, not just a single anomalous point), partly through integration of multiple sensor types, and partly through transparent reporting of confidence levels and coverage gaps in their products.

There is a fourth, subtler trap that experienced analysts discuss: over-detection sensitivity. A pixel-level change detection algorithm running globally and continuously will produce false alarms if the detection threshold is set too low. Agricultural fields undergoing routine tillage appear to show deforestation in some indices. Temporary water on flooded agricultural fields can appear to be a permanent new body of water. Shadow changes from different sun angles between image dates can mimic vegetation changes. The craft of building a reliable change-detection product involves not just sensitivity (detecting real changes) but also specificity (avoiding false alarms). These two goals pull in opposite directions, and every operational system involves a calibration between them.

Why Every Product Is a Time Series

Here is the most important conceptual shift this chapter offers: almost everything the geospatial industry calls a "product" today is not actually an image. It is a time-series engine with an alert layer on top.

Think about global deforestation alerts. Services like Global Forest Watch, RADD (Radar for Detecting Deforestation), and FORMA (Forest Monitoring for Action) do not work by taking two pictures and comparing them. They work by ingesting every available satellite pass for every forested pixel on Earth, updating a continuous model of each pixel's expected state, and generating an alert whenever a new observation deviates persistently from that model. The output a user sees on a map is the endpoint of a process involving thousands of images and a continuously running time-series engine.

Illegal fishing detection works the same way. Vessels in the ocean leave patterns in SAR and optical time series, predictable routes, vessel-specific signatures, and expected behavior in known fishing grounds at known seasons. AIS (Automatic Identification System) broadcasts let analysts track legal vessels, but vessels that turn off their AIS transponder and disappear from the broadcast record still appear in satellite imagery. By running a time series of SAR observations for a given ocean region and comparing observed vessel density against the pattern expected from AIS data, analysts can flag probable dark vessel activity, fishing or shipping that is deliberately concealing its position.

Infrastructure monitoring is another example. Tracking whether a country is building new airstrips, expanding port facilities, adding storage tanks, or constructing missile launch pads requires systematic comparison of high-resolution imagery across time. It is not one analyst looking at one picture. It is an automated time-series system flagging pixels that have changed from one landcover class to another, followed by human verification of the flagged objects. The analyst's job has shifted from "find the change" to "verify the automatically flagged change," a fundamentally different workflow.

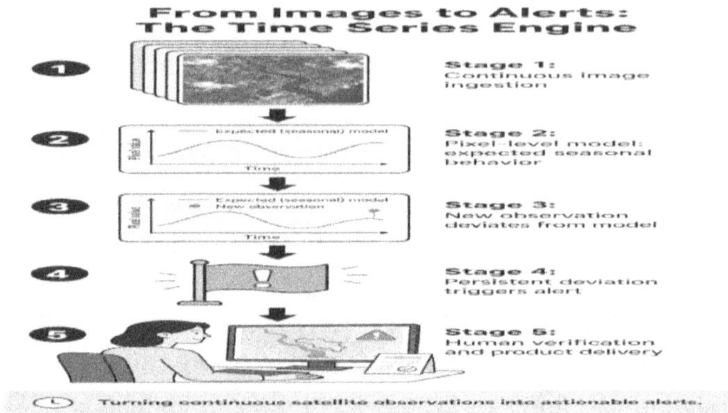

Diagram 5.12: From Images to Alerts: The Time Series Engine

Even agricultural products have made this shift. Early satellite-based crop yield forecasting services would produce annual or seasonal reports based on a limited number of image acquisitions. Modern systems are continuously updated: as each new Sentinel-2 pass completes, the crop model for every field in the monitoring area is updated with the new observation. Clients do not receive a seasonal report. They receive a continuously updated dashboard whose numbers reflect the latest available data.

Climate science products have the longest history with this approach because the science fundamentally requires it. Sea-surface temperature anomaly maps, which show where the ocean is warmer or cooler than the historical average for that time of year, are not snapshots. They are the output of a continuously running system that compares today's measurement at every pixel against a historical baseline model for that pixel and that time of year. The "anomaly" is defined relative to the time series baseline, not relative to any specific prior date. El Niño events, marine heat waves, and ice-extent anomalies are all defined and tracked through exactly this framework.

The shift from products as images to products as time-series engines has had significant commercial and operational implications. Building a time-series engine requires far more data infrastructure than producing a single image. It requires consistent archiving of every satellite pass, robust data pipelines, calibration systems to ensure that observations from different dates are radiometrically comparable, and algorithms that can run continuously and scalably at global coverage. Companies that have invested in this infrastructure have a durable advantage over those that produce one-off image analyses. The archive is the asset. The longer the time series runs and the better maintained the archive, the more valuable the product.

Takeaway

Change is what makes the planet legible over time. A single image is a record of one moment. A time series is a biography of a place. Learning to read those biographies, to separate the seasonal rhythms from the directional shifts, to know which indices amplify which events, and to recognize the traps that can fool even experienced analysts, is the core skill of modern Earth observation.

The concepts in this chapter connect to everything that follows. When you encounter a deforestation monitoring system, you now understand that it is a BFAST or CCDC-family algorithm running continuously on a stacked archive of images. When you see a news story about phenological shifts and earlier springs, you know that the evidence behind it comes from decades of NDVI time series extracted from Landsat and Sentinel archives. When a wildfire severity map is published within days of a fire's containment, you understand that it is an NBR differencing product, not a photograph.

You also now carry the healthy skepticism of an informed reader. When someone shows you a before-and-after comparison, you know to ask: which season were both images taken in? Are the dates matched to minimize phenological confounding? When someone reports a trend in vegetation, you know to ask: how was the sensor drift accounted for? Were the cloud gaps in the time series filled, and if so, how? These are the questions that separate a careful analysis from a compelling but misleading slide.

The satellite time series is one of the most powerful scientific instruments humanity has ever built, not because any single image is remarkable, but because the relentless accumulation of measurements, year after year, at consistent intervals, creates a record that outlasts any individual research project or government program. Somewhere in a data center right now, a satellite is completing its orbit and adding one more observation to a pixel time series that has been running for decades. The story of that place continues to be written, one overpass at a time.

6 Teaching Computers to See the Earth

Scenario

Daniela Reyes pulled up the vendor demo on a Tuesday morning and watched the map populate in real time. Red polygons blinked into existence across a 50,000-square-mile region, each one marking a new construction site that had appeared in satellite imagery over the previous week. The whole process took forty minutes. She sat with that number for a moment. Two analysts, working full time, used to spend a month doing the same job. They flagged roughly the same sites, but with gaps, especially in the regions at the edge of their assigned tiles where fatigue set in toward the end of a long shift.

The vendor's engineer walked her through the architecture without delving into the math. The model had been pretrained on hundreds of millions of satellite scenes from around the planet before it ever saw a single example from Daniela's region. It already understood rooftops, dirt, pavement, vegetation, shadows, and the particular texture of freshly graded earth. When the vendor's team fed it labeled examples of construction sites specific to her client's area of interest, the model adapted in hours rather than weeks. The engineer called it a foundation model. The term would stick in Daniela's vocabulary from that day forward.

She did not simply sign the contract and walk away. She pulled up three known false positives from the previous week's run, regions where the model had flagged what turned out to be bare agricultural fields cleared for the season rather than construction. She documented them carefully and sent them back to the vendor. Even a forty-minute pipeline, she told Marcus later, earns its keep only if a human is still looking at the edge cases. That combination of fast machine and watchful analyst is what this chapter is really about.

Why It Matters

Most readers encounter geospatial artificial intelligence in headlines first. "AI Spots Illegal Deforestation." "Satellite Model Counts Every Car in China." The headlines are real, and the capabilities behind them are genuinely impressive. But headlines do not tell you how the technology works, when it fails, or why the analyst sitting beside the screen still matters enormously. This chapter fills that gap.

You do not need to understand the mathematics of machine learning to become an informed consumer of geospatial AI. You do need to understand the ideas: where this technology came from, what kinds of problems it solves well, and where it stumbles. Whether you are evaluating a vendor pitch, interpreting a media story, or making a decision that depends on AI-derived imagery products, the concepts in this chapter will help you ask the right questions and resist both the hype and the fear that tend to travel together with any new technology.

The story of teaching computers to see the Earth is, in the end, a story about labor, data, and judgment. Machines supply the scale. Humans supply the context. The interesting part is understanding exactly where the line falls between the two.

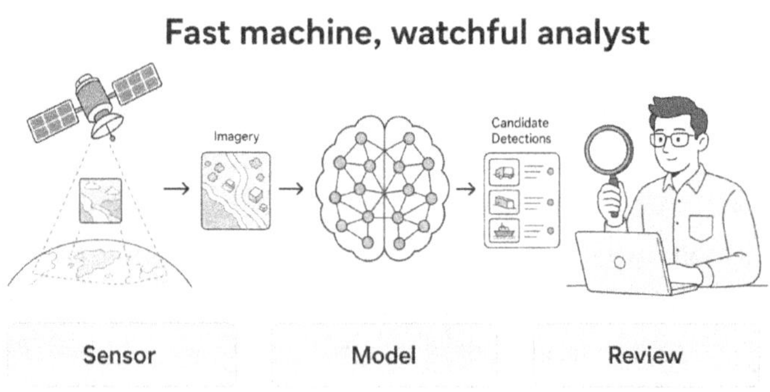

Diagram 6.1: The Human-Machine Partnership

Classical ML Came to Remote Sensing First

Before anyone used the phrase "deep learning" in a geospatial context, analysts were already using machine learning (ML) to classify land cover from satellite imagery. Machine learning, as a term, refers to any computational method that improves its performance by learning patterns from data rather than by following rules written by a programmer. The classical flavor of ML is not about neural networks or deep learning at all. It is about decision trees, random forests, and support vector machines. Understanding these older tools helps explain why the newer ones exist and why the older ones have not disappeared.

A random forest works something like a committee of experienced advisors who each examine the evidence independently and then vote. In a land-cover classification task, each member of the committee (called a decision tree) looks at a pixel's spectral values, the brightness values across different wavelength bands, and answers a series of yes-or-no questions. Is the near-infrared value above a certain threshold? Is the red band below another? The tree works through these questions until it reaches a leaf node, which is the equivalent of a final verdict: this pixel is cropland, or forest, or urban pavement. A random forest builds hundreds of these trees, each one trained on a slightly different random sample of the data, and combines their votes into a single prediction. The combination reduces the chance that any single bad tree corrupts the final answer.

Support vector machines (SVMs) take a different approach. An SVM tries to find the cleanest possible boundary between categories in a high-dimensional space. Imagine plotting every pixel in your image as a point in a space where each axis represents one spectral band. Cropland pixels cluster together in one region of that space. Urban pixels cluster in another. An SVM draws the widest possible gap between those clusters and uses that gap as the decision boundary for future classifications. SVMs became popular in remote sensing partly because they handle high-dimensional data gracefully and partly because they often produce reliable results from relatively small training sets.

These classical methods dominated satellite image analysis for roughly two decades, from the mid-1990s through the early 2010s. They powered global land-cover products like the MODIS Land Cover Type dataset and drove operational mapping programs for deforestation monitoring, crop area estimation, and urban expansion. They remain in production today at dozens of government agencies and commercial firms. Any honest survey of geospatial AI has to begin here, because it establishes a baseline: these tools are not glamorous, but they are proven, explainable, and far cheaper to run than their deep-learning successors.

Their limitations were real, though. Classical ML methods treat each pixel largely in isolation or within a very small neighborhood. They struggle with spatial context. A random forest classifying pixels does not naturally learn that a rectangular cluster of similar pixels surrounded by a road network is almost certainly a building. It sees brightness values, not shapes. It sees statistics, not structure. That gap opened the door to the next wave.

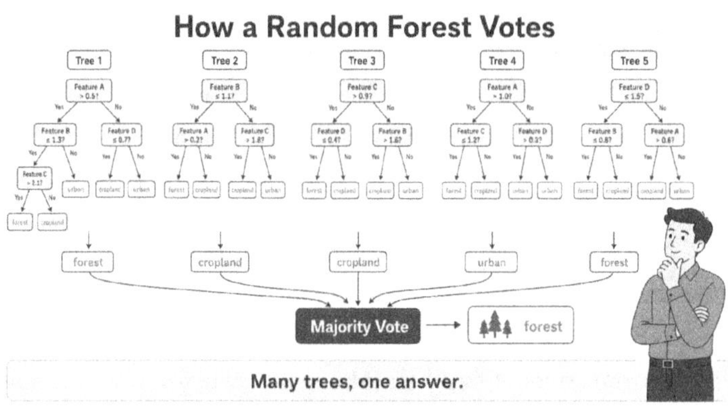

Diagram 6.2: How a Random Forest Votes

The teacher-student analogy from earlier in this book applies here too. In classical ML, you teach a committee of advisors by giving them labeled examples and letting them learn the rules. The committee is good at spotting statistical regularities. It is not so good at understanding composition, the relationship between parts of a scene rather than properties of individual points. That need for compositional understanding is exactly what the next generation of tools was designed to address.

Convolutional Neural Networks: The Second Wave

In 2012, a neural network called AlexNet won the annual ImageNet image recognition competition by a margin that shocked the research community. It did not just win narrowly. It won by such a margin that the entire field of computer vision effectively pivoted overnight. The technique behind AlexNet, a convolutional neural network (CNN, pronounced "see-en-en"), had existed since the late 1980s. Still, the combination of larger datasets, faster graphics processors, and better training tricks had finally made it work at scale.

A convolutional neural network is, at its core, a pattern recognizer that learns to detect structure in images by scanning for features at multiple scales. The name comes from a mathematical operation called convolution, but the intuition is simpler than the math. Think of the human visual cortex. When you look at a photograph of a building, your brain does not evaluate the whole image at once. It starts by detecting edges and lines, then combines them into shapes, and finally combines shapes into objects. The lower levels of your visual processing handle the details, and the higher levels handle the meaning. A CNN does something structurally similar. Its early layers learn to detect edges and color gradients. Its middle layers learn to combine edges into textures and textures into shapes. Its final layers learn to combine shapes into objects like rooftops, vehicles, and roads.

This layered, hierarchical approach is what made CNNs dramatically more powerful than classical ML for image-based tasks. A random forest sees pixels as isolated measurements. A CNN sees an image as a composition of nested patterns. That compositional understanding unlocked capabilities that had been out of reach for the older tools.

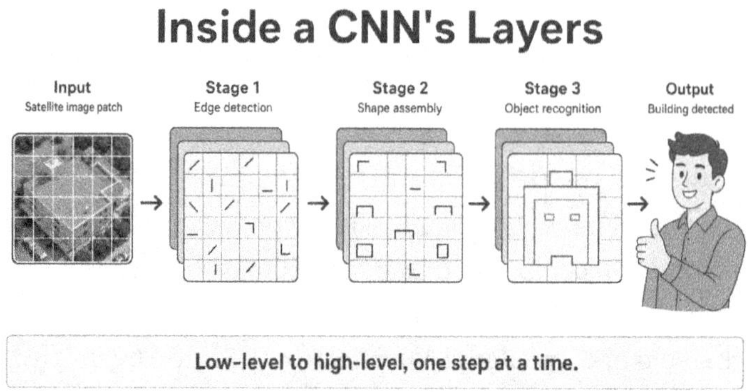

Diagram 6.3: Inside a CNN's Layers

Between roughly 2015 and 2022, CNN-based models reshaped what was possible in geospatial analysis. Building footprint extraction moved from a painstaking manual task to a semi-automated one. Road network detection became accurate enough to update map databases at national scale. Ship counting in ports could be done simultaneously across hundreds of harbors. The Microsoft Buildings dataset, which covers the contiguous United States and multiple other countries, contains machine-generated building polygons produced almost entirely by CNN-based models. The SpaceNet benchmark competitions, which we will return to later in this chapter, showed teams achieving accuracy levels unimaginable in the random-forest era.

CNNs also came with a new vocabulary that readers of geospatial AI reports should know. Semantic segmentation is the task of labeling every pixel in an image with a category, such as building, road, water, or vegetation. Object detection is the task of drawing a bounding box around every instance of a target class: every vehicle, every airplane, every shipping container. Instance segmentation combines both, labeling every pixel and distinguishing individual object instances from one another. When a vendor tells you their model does "semantic segmentation at 30-centimeter resolution," they are describing a CNN-based system that classifies every pixel in high-resolution imagery into a land-cover category.

CNNs were not without constraints. They required large amounts of labeled training data, which meant expensive human annotation. They were often brittle outside the geographic regions where they were trained. A building-detection CNN trained on North American suburbs would perform poorly on the densely packed, irregularly shaped structures found in many African and Asian cities. Each new task generally required its own dataset and its own round of training. And while CNNs were good at recognizing patterns within a fixed spatial context, they struggled with tasks that required understanding change over time or integrating information across very different image types. Those limitations set the stage for what came next.

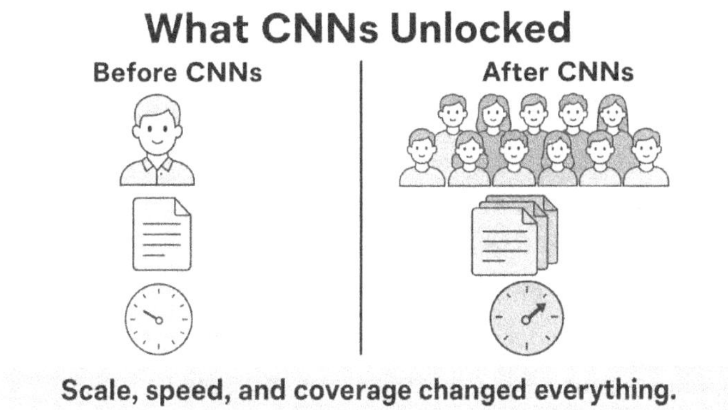

Diagram 6.4: What CNNs Unlocked

Foundation Models for Earth Observation

Around 2020, a shift began in the broader field of artificial intelligence that would eventually transform geospatial analysis the same way CNNs had transformed it half a decade earlier. Researchers developed the concept of the foundation model: a very large model pretrained on an enormous, diverse dataset, capable of being adapted to a wide range of specific tasks with relatively little additional training. The term "foundation model" is not a technical specification. It is a conceptual description. These models are called foundation models because they serve as a starting point, a foundation, for building task-specific applications on top.

The analogy that makes this intuitive is the experienced generalist. Imagine hiring a doctor who has spent twenty years working in emergency rooms, rural clinics, and research hospitals across six continents. That doctor has seen an enormous range of conditions, developed broad pattern recognition, and built a deep intuitive library of what the human body looks like in health and illness. When you assign that doctor to specialize in cardiac surgery, they do not start from scratch. They already know anatomy, physiology, pharmacology, and how to read a patient's demeanor. They need additional specialized training, but far less than a new graduate would. A foundation model is that experienced doctor. It has already processed the equivalent of twenty years of general experience before it encounters your specific problem.

For geospatial analysis, the relevant foundation models are trained not on text or general photographs but on vast quantities of satellite and aerial imagery. Prithvi is a geospatial foundation model developed by NASA and IBM, pretrained on years of Harmonized Landsat Sentinel-2 imagery covering the entire land surface of the Earth. Clay is an open-source geospatial foundation model designed to support a wide range of Earth observation tasks and trained on multiple sensor types. SatMAE is a model that uses masked autoencoding to learn rich representations of satellite imagery by training on unlabeled images alone. Major TOM is a large-scale dataset and an associated model ecosystem designed to provide the research community with a shared foundation for Earth observation AI development. These models are named here because readers will encounter them in vendor pitches, research papers, and media coverage. None of them requires mathematical explanation to understand their significance.

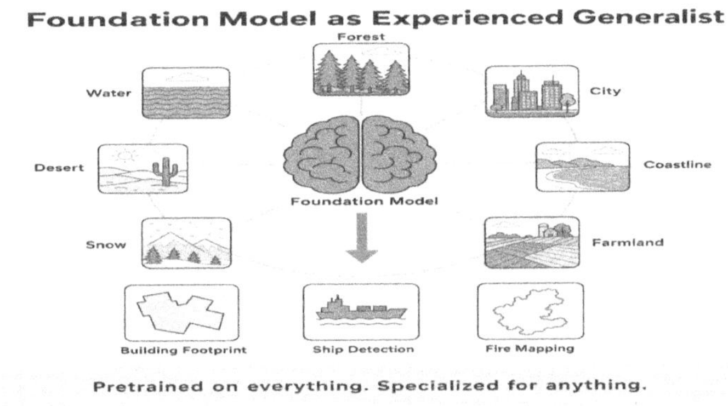

Diagram 6.5: Foundation Model as Experienced Generalist

The economics of this shift are significant. Before foundation models, building a geospatial AI product for a new region or task typically required gathering hundreds or thousands of labeled training examples specific to that task, training a model from scratch or heavily fine-tuning it, and validating extensively in the target geography. That process could take months and cost tens of thousands of dollars in labeling costs alone. With a foundation model, the starting point is already rich. Fine-tuning on a small number of labeled examples from the new task can produce competitive results in days rather than months. Daniela's vendor in the opening scenario was not performing magic. They were leveraging this shift in starting conditions.

Foundation models also generalize better across sensor types than their predecessors did. A well-trained geospatial foundation model can often transfer knowledge from optical imagery tasks to multispectral or synthetic aperture radar tasks with appropriate adaptation. This cross-sensor flexibility is genuinely new. It reflects that these models have internalized representations of the Earth that go beyond any single spectral configuration, much like an experienced human analyst who has worked with many sensor types can bring contextual understanding to an unfamiliar image product.

The limitations matter too, and they will be addressed in detail later in this chapter. Foundation models are not omniscient. They have never seen certain geographies. They have strong biases toward the regions and seasons most represented in their training data. They can be confidently wrong. Understanding their power requires understanding those limits in equal measure.

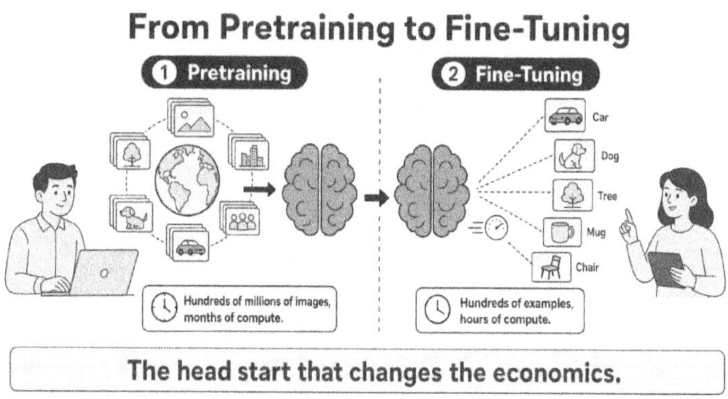

Diagram 6.6: From Pretraining to Fine-Tuning

Training Data: The Secret Ingredient

There is a phrase that data scientists repeat often enough that it has become something close to a law: a model is only as good as its training data. In geospatial AI, this principle has sharper edges than in almost any other field. Satellite imagery of the Earth is abundant, but labeled satellite imagery is not. Labeling means a human being examines an image and annotates it: draws a polygon around every building, marks the boundary between cropland and forest, identifies every vehicle in a parking lot. That work is slow, painstaking, and expensive. It also requires domain expertise. Not just any crowdworker can reliably annotate the difference between an irrigated alfalfa field and a dry pasture in a multispectral image.

The geospatial AI community has invested significant effort in building shared public datasets that reduce this bottleneck. xView is a dataset of satellite imagery labeled for object detection that covers more than 1,400 square kilometers and contains more than one million labeled objects across 60 classes, from buildings and small vehicles to large aircraft and maritime vessels. The Defense Innovation Unit released it to support research into AI-assisted overhead imagery analysis. SpaceNet is a collection of satellite imagery challenges sponsored by a consortium of commercial and government partners that has produced labeled datasets for building footprint detection, road network extraction, and off-nadir building detection. BigEarthNet is a large-scale benchmark dataset of Sentinel-2 multispectral imagery patches labeled with multiple land-use categories derived from European CORINE land-cover maps. These are not the only public datasets, but they are widely used, and readers who encounter them in vendor documentation or research papers now have a point of reference.

What these datasets share is that they represent particular geographies, seasons, and imaging conditions. xView imagery is largely North American. BigEarthNet is European. SpaceNet challenges have included diverse urban environments, but coverage is not uniform. When a model is trained primarily on these datasets, it absorbs the particular visual character of those places. North American building rooftops look different from those in South Asia. European agricultural parcels have a different shape and regularity than smallholder fields in sub-Saharan Africa. A model trained on one visual world will carry those expectations into every new image it analyzes.

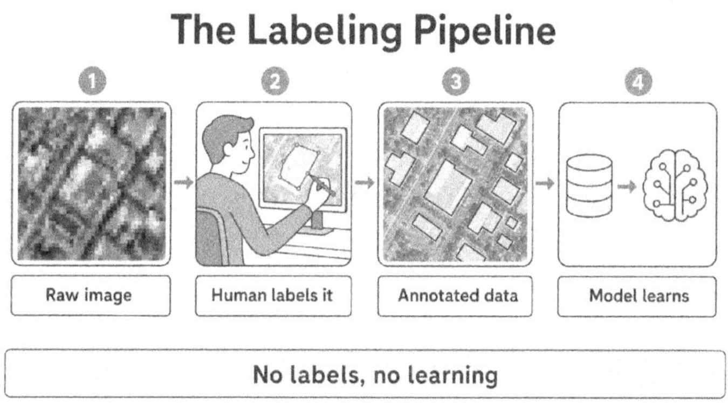

Diagram 6.7: The Labeling Pipeline

Analysts who work with geospatial AI models spend far more time on data preparation than on model design. This surprises many people who associate AI development with writing complex algorithms. In practice, the algorithm (especially with foundation models) is often already chosen. The bottleneck is the data: collecting it, cleaning it, verifying that the labels are accurate and consistent, handling edge cases where even experienced annotators disagree, and ensuring the training set reflects the diversity of conditions the deployed model will encounter. A geospatial vendor who says their model took 3 days to train is often describing only the final training step. The preceding months of data curation are the real labor.

The consistency problem deserves particular attention. When multiple people label the same image, they often disagree at the boundaries. Is this polygon a building or a shed? Does this pixel belong to the road or the sidewalk? Does this tree canopy count as one polygon or three? These disagreements introduce noise into the training data. Models trained on inconsistent labels develop inconsistent behavior. One widely used measure of labeling quality is inter-annotator agreement, which quantifies how often different human labelers assign the same labels to the same examples. High inter-annotator agreement is expensive to achieve but essential for training reliable models.

The future of training data for geospatial AI runs in two directions simultaneously. On one side, automated labeling techniques, including the use of foundation models to generate draft labels that humans then correct, are reducing the cost of annotation. On the other hand, the expansion of AI applications into new geographies and sensor types is creating a constant demand for labeled data in new domains. The two trends are racing each other. For now, the bottleneck remains real, which is why experienced geospatial AI practitioners spend so much of their time talking about data rather than algorithms.

How Accurate Is Geospatial AI, Really?

This is the question that vendor conversations tend to sidestep and that practitioners argue about at conferences. The short answer is: it depends, and the conditions it depends on are the most important part of the answer. Accuracy in geospatial AI is not a fixed number attached to a model. It is a function of what you are measuring, where the imagery was collected, when it was collected, and how the ground truth was established. Understanding this conditionality is one of the most practical skills this book can give you.

The most common accuracy metric in geospatial AI for polygon-level tasks is Intersection over Union, abbreviated as IoU (pronounced "eye-oh-you"). IoU measures how well a model's predicted polygon overlaps with the correct polygon drawn by a human expert. If the model draws a perfect polygon that exactly matches the human-drawn polygon, the IoU is 1.0. If the model's polygon shares only half the area of the correct polygon, the IoU is roughly 0.5. A threshold of 0.5 IoU is commonly used as the cutoff for calling a detection "correct." Higher thresholds, like 0.75 or 0.9, are more demanding and more realistic for precision applications.

Modern building-footprint models achieve IoU scores above 0.90 on benchmark tests that use North American suburban imagery. That number sounds impressive, and on those specific test sets, it is. But the word "benchmark" hides an important limitation. A benchmark is a standardized test, and standardized tests measure performance under controlled conditions. Performance in the field routinely differs from benchmark performance, sometimes by a large margin.

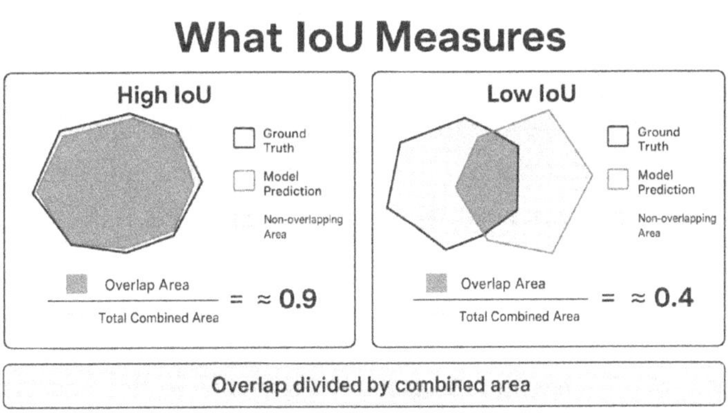

Diagram 6.8: What IoU Measures

The degradation patterns in geospatial AI accuracy follow predictable lines once you understand the training data story from the previous section. Building-footprint models that achieve 90-plus percent IoU on North American suburbs can drop to 60 or 70 percent on dense informal settlements in West African cities, where buildings are smaller, more tightly packed, share walls, and look spectrally different from what the model was trained on. Agricultural field-boundary models trained on European rectangular parcels perform poorly on the irregular smallholder plots common across much of Asia and sub-Saharan Africa. Road-detection models can confuse unpaved tracks and dry riverbeds in arid regions because their spectral signatures overlap and the models have seen few examples of either.

Season and illumination add another layer of variability. A crop-type classification model trained on summer imagery may fail spectacularly on winter imagery of the same fields, because bare winter fields look nothing like the dense green canopy the model learned to associate with its crop categories. A ship-detection model trained on open-ocean imagery may miss vessels docked in ports, surrounded by cranes and warehouses, because the visual context is completely different from that in its training set. These are not exotic edge cases. They are predictable failure modes that experienced practitioners anticipate and plan for.

The honest framing for accuracy in geospatial AI is not "this model achieves X percent accuracy" but rather "this model achieves X percent accuracy on this class of imagery, in this geographic region, under these seasonal conditions, relative to this definition of ground truth." That longer sentence is harder to put in a marketing slide, but it is the sentence that tells you whether the accuracy number is meaningful for your specific application.

Bias, Failure Modes, and Human Review

The word "bias" in machine learning has a technical meaning that overlaps only partly with the everyday meaning. In everyday speech, bias implies unfairness or deliberate distortion. In machine learning, bias refers to any systematic tendency of a model to err in a particular direction. Geospatial AI models carry biases that are almost always unintentional but often predictable, and understanding them is essential for anyone who relies on model outputs to make real decisions.

The most common source of bias in geospatial AI is geographic imbalance in training data. If a model learns primarily from imagery of one region, it will develop implicit expectations about what the world looks like, rooted in that region's particular appearance. Colors, textures, building styles, crop patterns, road widths, urban density, vegetation types: all of these vary enormously across the planet, and a model that has seen only a fraction of that variation will be surprised, and often wrong, when it encounters the rest.

A second important failure mode is confident wrongness. Classical statistical models often produce calibrated uncertainty, meaning that when they are unsure, they express that uncertainty in their output probabilities. Deep learning models, including CNNs and foundation models, are sometimes poorly calibrated in this regard. They can assign very high confidence scores to incorrect predictions. A model that says "building, 97 percent confident" about a large shadow patch is not lying. It is reporting its internal state accurately. But that internal state is wrong, and the high confidence number makes it easy for an analyst to skip the review step and act on the error. Calibration, the alignment between a model's stated confidence and its actual accuracy, is an active area of research for exactly this reason.

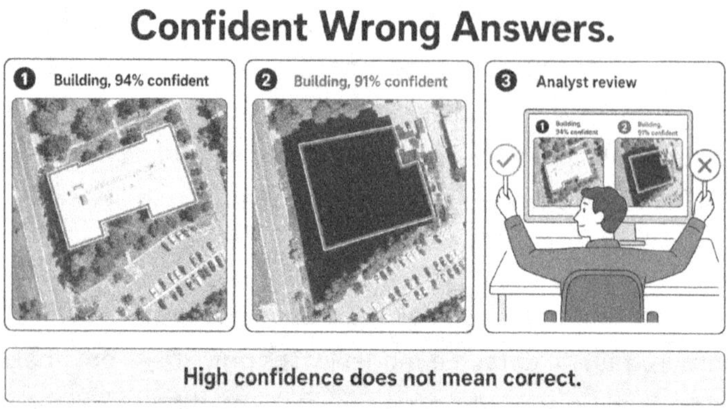

Diagram 6.9: Confident Wrong Answers

A third failure mode is adversarial confusion, which occurs when a scene contains elements not well represented in the training data and that closely resemble the target class, fooling the model. A gravel parking lot can look like an unpaved road or a dry riverbed depending on the imaging angle. A greenhouse roof can look like a solar array. A freshly burned field can look like bare soil or urban pavement. These are not exotic theoretical concerns. Practitioners encounter them regularly, and they require human review to resolve.

The concept of human-in-the-loop validation refers to the practice of keeping a human analyst in the decision-making chain even when a model produces outputs. It is not a fallback for when the model fails. It is a structural principle for designing responsible geospatial AI systems. The analyst is not there to rubber-stamp the model's outputs. The analyst is there to identify failure modes, provide feedback to improve future versions of the model, and accept accountability for decisions that affect people and resources.

The human-in-the-loop principle has both a quality argument and an ethical argument. The quality argument is practical: models that receive human feedback over time improve faster than models deployed in isolation. When an analyst flags a confident wrong answer, that flagged example becomes a training signal for the next model version. The system learns from its mistakes because a human identified the mistakes. The ethical argument is more fundamental: decisions that affect people's lives, whether they concern property boundaries, humanitarian aid distribution, or military targeting, should not be delegated entirely to automated systems that produce confident wrong answers. Human accountability is not a limitation of AI. It is a feature of responsible AI.

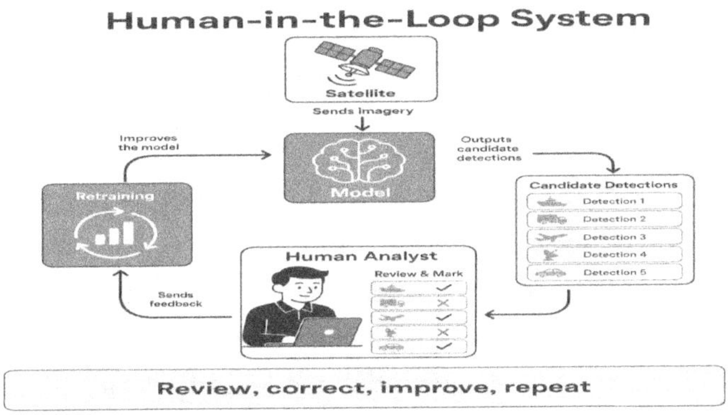

Diagram 6.10: Human-in-the-Loop System

Marcus asked Daniela once why the vendor's contract included a clause requiring a human analyst to review every flagged construction site before the result was delivered to the client. Daniela's answer was practical. The client was making infrastructure investment decisions based on those flagged sites. A false positive might send a field team to the wrong place. A false negative might cause them to miss a significant development. Neither the vendor nor Pleiades was willing to put their name on a fully autonomous pipeline for a decision of that consequence. The clause was not a concession. It was the whole point.

A Grounded Five-Year Forecast

Forecasting the trajectory of any technology is a high-confidence way to be at least partly wrong. The appropriate approach is not to hide behind that uncertainty but to reason carefully from what is already in motion, distinguish between what is nearly certain, what is plausible, and what is genuinely unknown, and be honest about the difference. This section tries to do exactly that.

What is nearly certain over the next five years is this: foundation models for Earth observation will proliferate and improve. The trajectory is already established. Computing costs for training large models have dropped by roughly a factor of 10 every 4 to 5 years over a sustained period. As inference costs fall, running a foundation model to analyze a satellite image becomes cheaper than running a classical ML pipeline. More open datasets will be released, more pretrained models will be shared by the research community, and more small vendors will be able to build competitive products on top of them without training from scratch.

Global coverage will also improve. One of the persistent frustrations in geospatial AI has been the uneven distribution of training data, with dense coverage in North America and Europe and sparse coverage in many other regions. New labeling initiatives, synthetic data generation, and transfer learning from foundation models are all contributing to filling those gaps. A building-detection model that fails on West African informal settlements today will perform much better in three to five years, not because the underlying architecture has changed, but because the training data has improved.

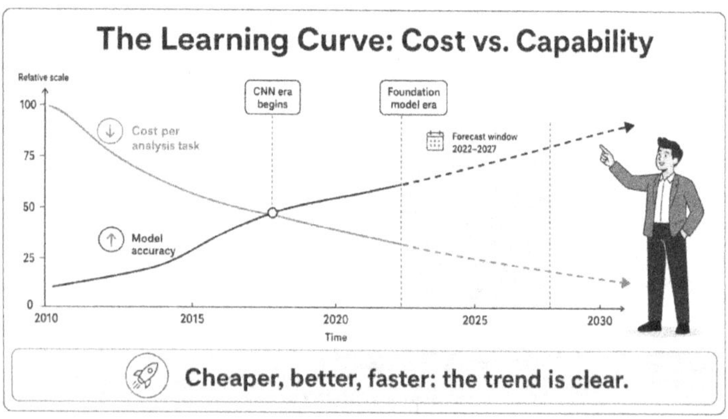

Diagram 6.11: The Learning Curve: Cost vs. Capability

What is plausible but not certain: the integration of geospatial AI with real-time sensor streams. Today, most geospatial AI applications operate on imagery that is hours or days old. As satellite revisit rates increase and as on-board processing capabilities improve (models running directly on the satellite rather than on the ground), near-real-time analysis becomes more feasible. The potential applications range from disaster response, where the first hours matter enormously, to dynamic monitoring of rapidly changing phenomena like wildfires and flood boundaries. But the infrastructure, both in orbit and on the ground, is not yet in place to make this routine, and the cost reductions needed to make it accessible to all potential users have not yet materialized.

Improved multimodal models, systems that can seamlessly combine optical imagery, synthetic aperture radar, hyperspectral data, and non-imagery signals like ground sensor readings or weather data, are also plausible within five years. The research foundations are being laid now. Whether the operational deployments will follow at the pace suggested by the research is a different question.

What is overhyped: fully autonomous geospatial decision-making without human review. This will not be standard operating practice for high-stakes applications within five years. The failure modes described earlier in this chapter are real, and the consequences of confident wrong answers in consequential domains are significant enough that organizations will maintain human oversight. The goal is not to remove the analyst. The goal is to give the analyst better tools so that their judgment is applied to the most important decisions rather than to routine extraction tasks.

What is genuinely uncertain: the regulatory and ethical landscape. Questions about who owns the imagery of the Earth, what privacy protections apply to the continuous monitoring of populated areas, and who is accountable when a geospatial AI system produces a harmful error remain unsettled. Different national jurisdictions are moving in different directions. The answers will shape both what geospatial AI can legally do and what it will be trusted to do, and they will emerge from policy processes that are still in early stages.

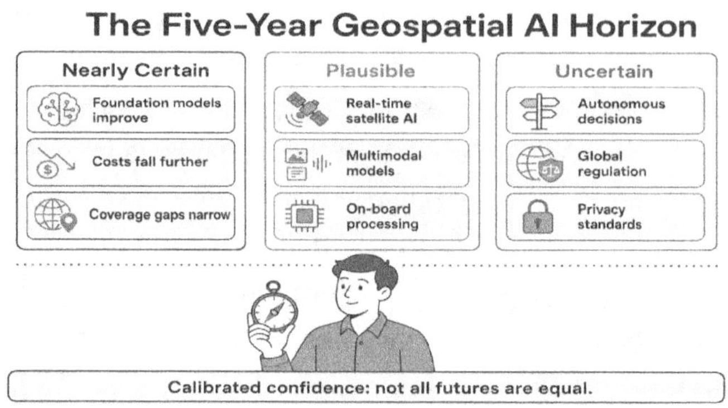

Diagram 6.12: The Five-Year Geospatial AI Horizon

The readers who will navigate this landscape most effectively are not the ones who read every technical paper. They are the ones who understand the structural principles well enough to ask vendors good questions, interpret media coverage accurately, and know when a result warrants more scrutiny before it drives a decision. Those are exactly the skills this chapter has been building.

Takeaway

Teaching computers to see the Earth has been a decades-long project that moved through distinct phases: classical machine learning, convolutional neural networks, and now foundation models. Each phase expanded what was possible and inherited the limitations of what came before. Understanding the progression is more valuable than knowing the name of any particular model, because the progression tells you something stable: this technology improves by learning from data, and its blind spots are always rooted in the gaps in that data.

The honest picture is more nuanced than either the optimist's or the pessimist's version. Modern geospatial AI is genuinely capable of things that would have seemed miraculous twenty years ago. It can count ships in every major port simultaneously. It can detect changes in building density across entire continents. It can flag potential infrastructure development in regions where no ground surveys have been conducted for years. These are real capabilities that are changing how analysts work, how resources are allocated, and how decisions about the physical world are made.

At the same time, geospatial AI fails in predictable ways, mostly in the places and seasons it has not seen before. It produces confident wrong answers. It inherits the biases of its training data. And in the most consequential applications, it is not yet trusted to operate without a human in the decision loop. That trust is being built incrementally, with feedback, validation, and transparency, which is exactly how trust in any powerful tool should be built.

When you encounter a headline about AI and satellites, you now have a framework for asking the right questions. What was the model trained on? Has it been validated in this geography? Was a human analyst involved in verifying the outputs? What does the error rate look like outside the benchmark test? Those questions do not require a computer science degree. They require the kind of informed skepticism that this chapter was designed to give you.

7 From Farms to Cities to Climate

Scenario

It is two-thirty on a Tuesday afternoon at Pleiades Geospatial Services, and the open-plan floor sounds like three different industries happening at once. Over by the windows, Priya Bhatt has a corn grower in Iowa on a video call. She is walking him through a set of false-color maps covering his 3,200-acre operation, pointing to a wedge of deep red near the southeast quarter section. That red means low chlorophyll. The crop has been under drought stress for 12 days. The grower says he knew something was off in that corner but could not see it from the road. Priya pulls up a second map showing soil-moisture estimates from a radar satellite pass at six in the morning, and the picture comes clear. The problem is concentrated in a zone where the soil drains faster than the rest. She tells him where to send his irrigation pivot tonight, and he writes it down.

Across the room, Marcus Okonkwo has a different kind of map filling both of his monitors. It is Phoenix in late July, and the city glows in a palette that runs from pale yellow to deep orange to a bruised red-violet. He is building an urban heat-island analysis for a city planner who wants to know which neighborhoods bear the greatest thermal burden during heat waves. The hot zones are not downtown office towers. They are older residential blocks with aging dark asphalt and very few trees. Marcus is annotating the map with cooling center locations, and the mismatch between where heat concentrates and where relief exists is stark.

At the back of the room, Dr. Amara Wells is preparing a methane-emissions audit for a European energy regulator. She has compiled a time series of satellite passes over a natural gas distribution network in northern Germany. The satellite, operated by GHGSat (Greenhouse Gas Satellite), carries a spectrometer tuned to methane's absorption wavelengths. Amara's screen shows point-source plumes above a handful of compressor stations. She is quantifying emission rates, timestamping each event, and cross-referencing with wind data to tell the regulator which facilities are leaking, how much, and for how long. Three employees, three clients, three completely different problems, and one underlying technology stack built from satellites, sensors, spectra, and time series. That is what this chapter is about.

Why It Matters

Most people think of satellite imagery as something used by governments, militaries, or scientists tracking glaciers. That picture is decades out of date. Today, a corn farmer adjusting an irrigation schedule, a city planner deciding where to plant trees, a hedge fund pricing a futures position, an insurance company writing a wildfire policy, a humanitarian organization mapping roads after an earthquake, and an energy regulator auditing methane leaks from a pipeline network are all drawing on the same fundamental technology. The scene at Pleiades is not a futuristic fantasy. It is Tuesday afternoon in a growing industry.

This chapter surveys that breadth. It is deliberately wide rather than deep. If you have been reading since Chapter 1, you already understand how sensors capture reflected and emitted energy, how time series reveal change, how spectral signatures distinguish one material from another, and how geospatial AI turns images into decisions. What this chapter shows you is where all of that lands in the real world. By the end, you will have a working map of the industries that depend on satellite data and a portable mental model that will let you recognize the same pattern wherever you encounter geospatial intelligence next.

Precision Agriculture and the Digital Farm

Farming has always been an exercise in managing uncertainty. Rain, heat, pests, soil variability: a grower makes thousands of decisions every season against a backdrop of incomplete information. For most of agricultural history, the information available was what you could see from the cab of a tractor or feel by walking a row. That is a narrow window onto a complex system. A typical Iowa corn field of several thousand acres contains more soil variability, more micro-drainage patterns, and more pest-pressure gradients than any one person can observe on foot.

Remote sensing changed the information geometry of farming over several decades, beginning with the Landsat program in the 1970s. Researchers noticed that healthy green vegetation reflected a great deal of near-infrared light and unhealthy or sparse vegetation reflected far less. That observation was eventually formalized into an index called NDVI, which stands for Normalized Difference Vegetation Index. NDVI compares the near-infrared reflectance of a pixel with its red reflectance and yields a value between -1 and 1. Healthy, dense, chlorophyll-rich crops score high. Bare soil, stressed crops, and water score low. The formula is simple. The insight it encodes is profound.

Priya has explained NDVI to growers hundreds of times and uses the same analogy each time. Think of a swimming pool light versus a reading lamp. A healthy leaf under near-infrared illumination blazes like a pool light seen from underwater. Stress the plant, withhold water, infect it with a fungus, and that light dims. NDVI is just a way of measuring how bright or dim that light is across every pixel in the image. Map those values across an entire field, and you get a picture of plant health at a resolution no human scouting team could produce in a day.

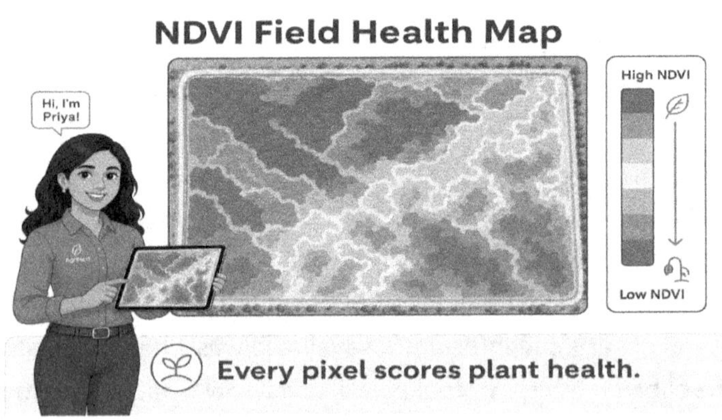

Diagram 7.1: NDVI Field Health Map

Modern precision agriculture layers multiple data sources onto the NDVI foundation. Thermal imagery, which measures the heat a crop surface emits rather than the light it reflects, reveals a different kind of stress. A well-watered plant stays cooler than its surroundings because it is transpiring: pulling water up from roots and releasing it through tiny pores in its leaves. When water is withheld, transpiration slows, and canopy temperature rises. Growers who layer thermal maps on top of NDVI maps get a two-dimensional picture of stress that separates nutrient problems from water problems, because each produces a different combination of low NDVI and elevated temperature.

Soil moisture is a third data layer, and this is where SAR (Synthetic Aperture Radar) earns its place in precision agriculture. SAR sends its own microwave pulses toward the surface and records how much energy bounces back. Wet soil and dry soil have different electrical properties, so they scatter radar energy differently. SAR sensors operating at longer wavelengths can even penetrate a plant canopy to interrogate the soil beneath it, making them useful on overcast mornings and in the rainy tropics where optical satellites may not see clear ground for weeks. For Priya's Iowa grower, the SAR pass at dawn confirmed the drainage pattern she suspected: the southeast zone was fast-draining sandy loam. NDVI confirmed the consequence. Together they pointed to the irrigation prescription.

Yield prediction models extend this logic into the future. By analyzing NDVI trajectories across an entire growing season for hundreds of fields and pairing those trajectories with historical yield records, machine-learning models can estimate end-of-season crop yields months before harvest. The USDA (United States Department of Agriculture) uses satellite-derived crop condition assessments as inputs to its official yield forecasts, which move global commodity markets when they are released. At the field level, a grower who receives a yield forecast in early August can adjust inputs, negotiate forward contracts at favorable prices, or shift harvesting logistics before the crop is in the ground.

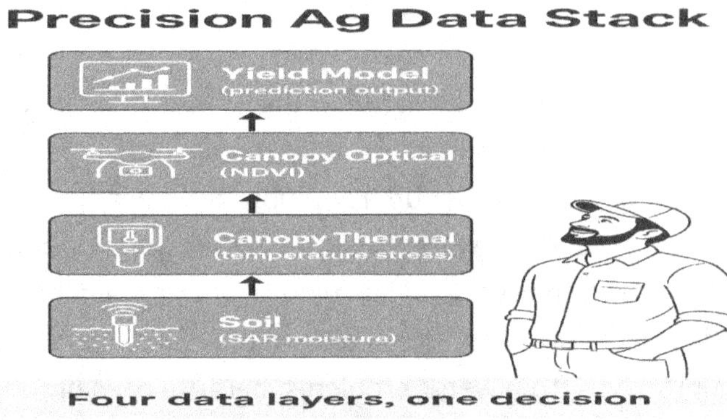

Diagram 7.2: Precision Ag Data Stack

The change goes beyond individual fields. The GEOGLAM (Group on Earth Observations Global Agricultural Monitoring) initiative aggregates satellite-derived crop assessments across all major growing regions worldwide to provide early warning of potential food shortfalls. When a drought in the Black Sea wheat belt appears in NDVI maps six weeks before harvest, policymakers have time to prepare import plans, activate emergency reserves, or communicate with vulnerable populations. The satellite data does not prevent droughts. It removes the surprise.

The economics have also shifted. Until around 2015, fine-resolution NDVI data was expensive to access. Planet Labs and other commercial operators launched fleets of small satellites that now image nearly every agricultural region daily. The European Space Agency made Sentinel-2 imagery freely available worldwide at ten-meter resolution. A farmer in Iowa or a cooperative in Kenya can access more timely, higher-resolution vegetation data than a government researcher could obtain a decade ago for any price. Priya's team at Pleiades is one piece of an ecosystem that delivers actionable maps to a farmer's phone, built on processing pipelines that run nightly and would have required a week of specialist time in 2010.

Urban Heat Islands and the City We Live In

Put two identical houses side by side. Give one a dark asphalt shingle roof and a concrete driveway. Give the other a white reflective roof, a grass yard, and a row of shade trees. On a summer afternoon, the air temperature two feet above the concrete might be ten degrees Fahrenheit higher than the air above the grass. Scale that difference to an entire city district, and you have the urban heat island effect.

Cities are hotter than their surrounding countryside for a straightforward physical reason. Urban surfaces, such as asphalt roads, dark rooftops, parking lots, and glass-and-concrete facades, absorb more solar energy and emit more of it back as heat than vegetation-covered surfaces. Trees and grass also cool their surroundings through evapotranspiration, the same process that cools a sweating athlete. Replace vegetation with impervious surfaces, add the waste heat from air conditioners and vehicles, and you can add two to five degrees Celsius to the ambient temperature of a dense urban core relative to a forested suburb just ten miles away.

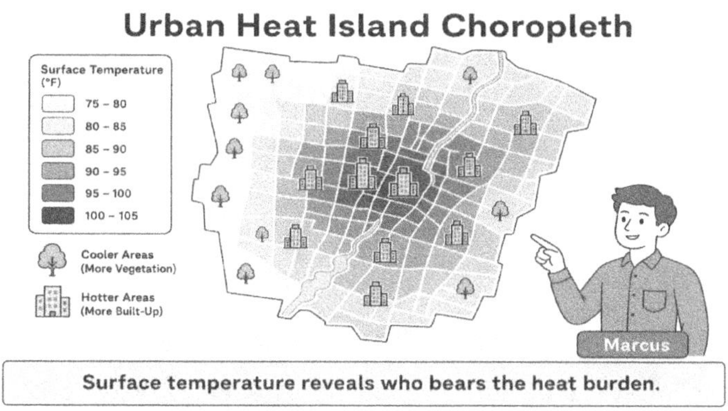

Diagram 7.3: Urban Heat Island Choropleth

Thermal satellite data makes the heat island visible in a way no network of street-level thermometers could. The Landsat thermal band and the ECOSTRESS instrument aboard the International Space Station both measure land surface temperature at thirty-meter to seventy-meter resolution, fine enough to distinguish a sun-exposed parking lot from an adjacent park a block away. When those maps are overlaid with neighborhood boundaries, demographic data, and tree-canopy coverage, a city planner can identify with precision which districts are most thermally vulnerable, which residents are most exposed, and which interventions would yield the greatest cooling benefit.

The Phoenix analysis Marcus was preparing illustrates how the data informs policy. When thermal maps are overlaid on income data, a consistent pattern emerges across many cities: lower-income neighborhoods tend to be hotter because they have less tree canopy, more impervious surfaces, and older building stock. This disparity is now recognized as an environmental justice issue, and satellite thermal data has given advocates and planners a quantitative tool to document it and make the case for intervention.

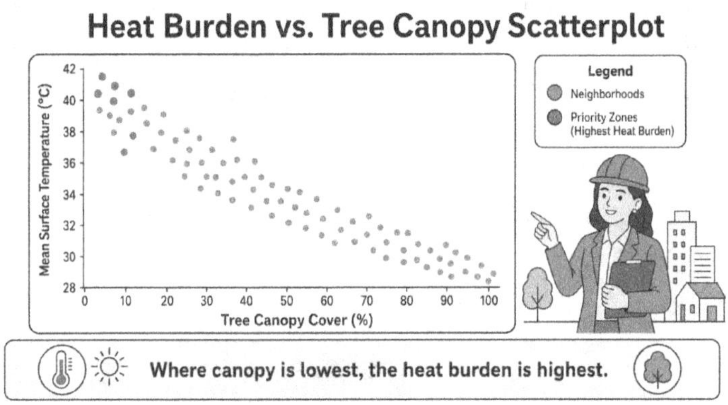

Diagram 7.4: Heat Burden vs. Tree Canopy Scatterplot

Tree-planting programs informed by satellite heat maps are already operating in several US cities, including Los Angeles, New York, Philadelphia, and Phoenix. The approach is systematic rather than aesthetic: identify the hottest, most densely populated, least-tree-covered zones, rank them by heat exposure and population vulnerability, and target planting there first. Satellite-derived canopy coverage data, often produced from high-resolution multispectral imagery or LiDAR (Light Detection and Ranging) point clouds, tell planners not just where trees are but also how much shade they cast, something impossible to derive from aerial photos alone.

Beyond heat, multispectral imagery combined with machine-learning classifiers can produce detailed land-cover maps of a city, distinguishing rooftops from roads, grass from shrubs, pools from ponds. These maps feed urban energy models, stormwater models, and flood risk models, because the distribution of impervious surfaces determines how quickly streets fill with water when a culvert overtops. Paris has used satellite-derived heat and vegetation maps to guide its Plan Canopée, targeting 170,000 tree plantings with priority zones determined by satellite analysis rather than aesthetics alone. The satellite data does not replace landscape architects and urban ecologists. It sharpens the targeting so that limited budgets go to the places that need them most.

Climate Science and the Planet's Slow Signals

Some changes on Earth happen fast enough to see from day to day: a wildfire, a flood, a storm track. Others happen on timescales of years to decades. They are real, large in magnitude, and profoundly consequential, but invisible to a single snapshot. You cannot see sea level rise in a photograph. You cannot see ice mass loss or permafrost thaw in Tuesday's image. You need a long, consistent, spatially comprehensive record of the same measurements, taken the same way, for years on end. That is exactly what satellite programs have been quietly building since the early 1970s.

The Landsat archive is the longest continuous land-surface remote sensing record in existence. Every image acquired since the first Landsat launched in 1972 has been preserved and is freely available through the USGS (United States Geological Survey) Earth Explorer portal. Researchers today can compare forest cover in the Brazilian Amazon in 1975 with current cover using imagery collected along the same orbital path with sensors designed for consistency across generations. The archive is not just scientifically valuable. It is irreplaceable. You cannot go back in time to reimagine a forest that was cleared before satellite coverage began.

Ice mass loss is one of the most consequential signals in the climate record, and remote sensing is the only tool capable of tracking it globally. The GRACE mission (Gravity Recovery and Climate Experiment) and its successor, GRACE-FO (Follow-On), measure tiny fluctuations in the Earth's gravitational field caused by changes in the mass of ice and water beneath the satellites. When the Greenland ice sheet loses ice, the gravitational pull at that location weakens by a measurable, extremely small amount. GRACE detects this by tracking the separation between two satellites flying in formation 220 kilometers apart. When one passes over a mass anomaly before the other, the gravitational gradient stretches or compresses their separation by a few microns. That signal, parsed with careful modeling, produces estimates of ice mass change at continental scale with a precision no ground-based network could achieve.

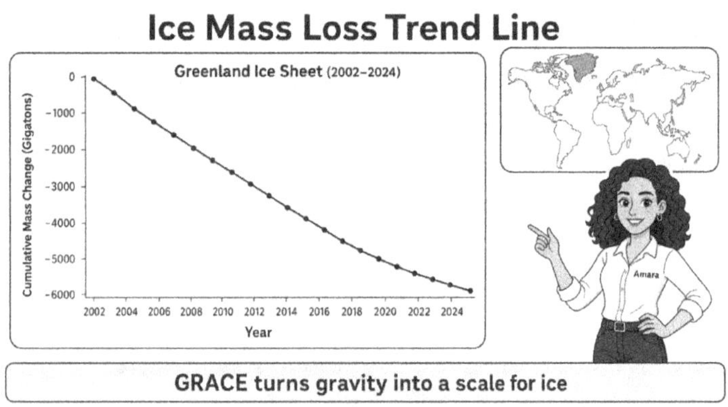

Diagram 7.5: Ice Mass Loss Trend Line

The measurements that follow are sobering without being sensational. Greenland has been losing ice at an accelerating rate over the decades. Ice shelves that act as buttresses holding back the flow of land ice have been thinning. These measurements come from satellites: GRACE for mass balance, radar altimeters for surface elevation, SAR interferometry for ice velocity, and optical sensors for calving-front positions. The satellite record does not predict what happens next. That requires climate models. But models cannot be calibrated without the observations satellites provide.

Deforestation and permafrost thaw are two other slow signals that satellite time series have clarified. Global Forest Watch, built on Landsat and maintained by the World Resources Institute, now reports tree cover loss at nearly real-time cadence. When tropical forest is cleared, the change appears in satellite imagery within days of the next cloud-free pass. Permafrost, ground continuously frozen for at least two years and covering roughly a quarter of the Northern Hemisphere's land area, releases carbon as it thaws and causes the surface above it to subside. Researchers use optical and radar time series to track the expansion of thermokarst features (pits, ponds, and slumps) across Siberia, Alaska, and Canada, using them as proxies for the pace of degradation.

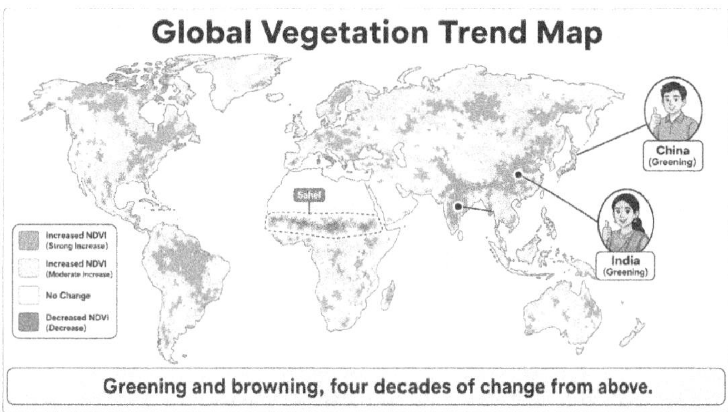

Diagram 7.6: Global Vegetation Trend Map

Ocean surface temperature is another planetary signal measured primarily from satellites. The MODIS (Moderate Resolution Imaging Spectroradiometer) instruments aboard NASA's Terra and Aqua satellites have been measuring sea surface temperature at one-kilometer resolution since 2000. The record reveals El Niño and La Niña cycles, documents the warming of the North Atlantic, and provides the input data for hurricane intensity forecasts. When a hurricane approaches an unusually warm patch of ocean, intensity models that incorporate satellite-derived sea surface temperature give forecasters a few more hours of warning before rapid intensification occurs. These applications share a single underlying truth: satellites are not optional extras for climate science. They are the observing backbone.

Insurance, Finance, and Risk

The insurance industry runs on risk models. For most of the twentieth century, those models relied on historical loss data, weather records, and actuarial tables. The models worked reasonably well when the future looked like the past. As climate change has altered the frequency and severity of extreme weather events, historical tables have become less reliable. Satellite data has stepped into that gap.

A property and casualty insurer writing policies in the wildland-urban interface (the zone where houses meet fire-prone vegetation) faces a difficult problem. Two houses on the same street can have very different fire risk depending on roofing material, surrounding vegetation, land slope, and how prevailing winds funnel through the local terrain. All of those factors are now measurable from satellite and aerial imagery. Vegetation type and density from multispectral classification. Roofing material from high-resolution optical analysis. Terrain from digital elevation models derived from radar or LiDAR. Insurers have built risk-scoring systems that produce parcel-level estimates far more granular than anything previously available.

After a wildfire or hurricane, satellite-derived damage assessments provide another critical input. Before an adjuster can set foot on damaged properties, often before roads are even passable, satellite imagery is being analyzed to classify the damage state of every structure in the affected area. Was this house destroyed? Partially damaged? Did the fire reach the wall but leave the structure standing? These classifications, produced by AI models trained on thousands of before-and-after image pairs, give insurers a rapid first inventory of loss that shapes reinsurance settlements, reserve requirements, and adjuster deployment priorities.

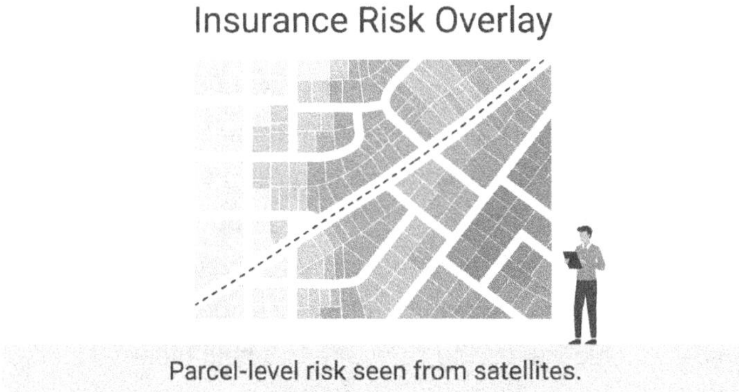

Diagram 7.7: Insurance Risk Overlay

The hedge-fund side of geospatial finance is less well known but has been operating for longer than most people realize. Alternative data providers discovered in the early 2000s that parking-lot car counts at major retailers, extracted from high-resolution satellite imagery a few days before quarterly earnings reports, gave a statistically useful signal about same-store sales. More cars correlate with more shoppers. Analysts task satellites with sampling retail locations before earnings announcements, count parked cars using automated object-detection algorithms, compare the counts to prior-year periods, and use the results to inform trading positions.

The same approach has been applied to oil-tank shadow analysis. Large crude oil storage tanks use floating-roof designs in which the roof rises and falls with the liquid level inside. A satellite image from directly overhead shows the tank's circular outline and a shadow cast by the roof lip. The depth of the shadow reveals the volume of oil inside. By monitoring hundreds of tanks globally on a regular cadence, commodity analysts can estimate global crude inventory builds and draws between official releases from the US Energy Information Administration and the International Energy Agency.

Flood risk modeling has gone through a similar transformation. Traditional flood maps in the United States, such as FEMA's (Federal Emergency Management Agency) National Flood Insurance Program maps, were derived from historical stream-gauge records and hydrological models, many of which are decades old. Satellite-derived elevation data at fine resolution, combined with satellite observation of actual flood extents from recent events, has enabled a new generation of flood models that are both more accurate and more current. Index-based agricultural insurance has been similarly transformed: index policies pay out when a satellite-derived metric such as NDVI falls below a threshold, removing the need for physical inspection and making it economically viable to provide coverage to smallholder farmers in Kenya, Ethiopia, India, and Bangladesh.

Disaster Response and Humanitarian Work

When a major earthquake strikes a city, a hurricane makes landfall, or a river crests its banks by twenty feet, the immediate problem for responders is information. Where are the damaged neighborhoods? Which roads are passable? Where are people trapped? In the immediate aftermath of a disaster, the people who need this information most urgently are often the same people who have lost communication infrastructure, can no longer trust their existing maps, and cannot get aircraft close enough to do a visual survey. Satellites can see through all of that.

Satellite-based disaster response has been formalized through the International Charter on Space and Major Disasters. The Charter is a voluntary agreement among major national space agencies and commercial satellite operators worldwide. When a qualifying disaster occurs and an authorized user (typically a national civil protection authority or United Nations agency) activates the Charter, member agencies task their satellites over the affected area and provide imagery at no cost. Data flows into the response operation within hours, bypassing commercial pricing and contracting that would otherwise slow access. The Charter has been activated hundreds of times since its founding in 2000, covering earthquakes, floods, volcanic eruptions, tropical cyclones, and wildfires on every populated continent.

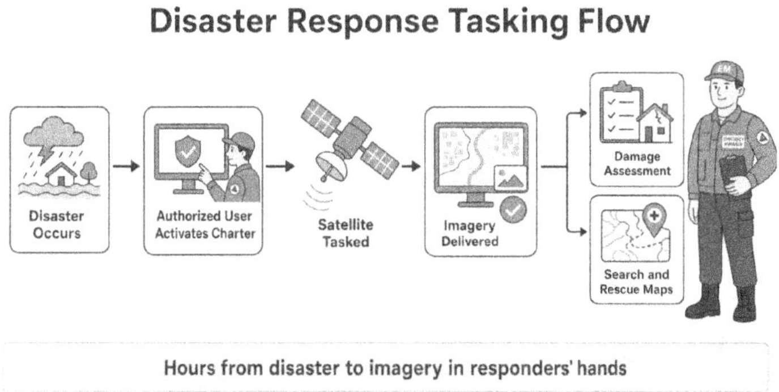

Diagram 7.8: Disaster Response Tasking Flow

A Charter activation typically produces two types of analysis. Rapid damage assessment compares pre-event and post-event imagery to classify the damage state of every building or infrastructure element in the scene. Change detection algorithms flag the differences; human analysts or AI models then interpret whether a change represents collapse, fire damage, flooding, or other effects. The output, typically a damage-density map, guides the deployment of search-and-rescue teams and medical resources to the zones with the heaviest losses first. Road passability assessment runs in parallel, identifying blockages visible from above and merging satellite observation with crowd-sourced field reports from people using smartphone apps on the ground.

HOT OSM (the Humanitarian OpenStreetMap Team) plays a critical role in this ecosystem. OpenStreetMap (OSM) is a collaboratively built global map, somewhat like a Wikipedia for geography. When a disaster strikes an area with limited map coverage, such as a secondary city in a low-income country, HOT OSM activates a rapid-mapping campaign. Volunteers around the world are shown high-resolution satellite imagery of the affected area and trace buildings, roads, paths, and water features into the shared database. Within 48 to 72 hours of activation, a community that previously existed as a blank space on digital maps can have a detailed street-level map guiding humanitarian responders to the right street, building, and door. The Haiti earthquake of 2010 established this model: within three days of activation, thousands of volunteers had produced a detailed map of Port-au-Prince that search-and-rescue teams from a dozen countries were using in the field.

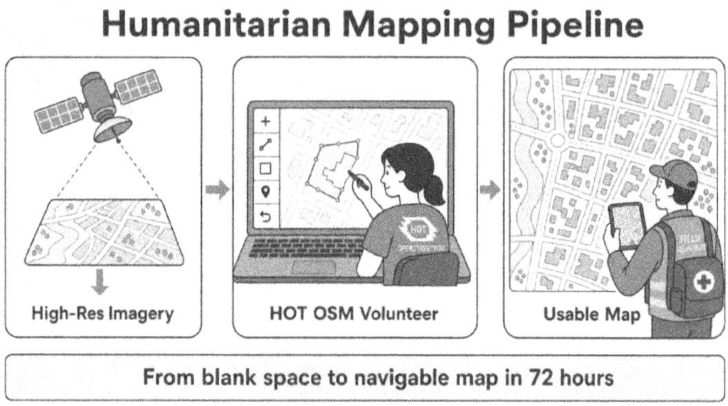

Diagram 7.9: Humanitarian Mapping Pipeline

Wildfire response relies on VIIRS (Visible Infrared Imaging Radiometer Suite) sensors aboard the Suomi NPP and NOAA-20 satellites, which provide twice-daily active fire detection globally at 375-meter resolution. Sentinel-2 and commercial high-resolution satellites provide detailed burn scar mapping within hours of a fire passing. During floods, SAR is particularly valuable because radar can see through clouds and operates at night. Water surfaces appear very dark in radar imagery, making flood extents easy to delineate. The Copernicus Emergency Management Service in Europe and SERVIR (a joint NASA and USAID program) provide near-real-time flood extent maps to national and local emergency managers during major events, showing which roads are cut off, which agricultural areas are inundated, and where floodwater is receding.

The humanitarian applications of geospatial intelligence share a common ethical dimension. The people most dependent on these tools in a crisis are also those with the least agency to influence the systems that produce them. An earthquake survivor in a poorly mapped city cannot request better satellite coverage. They rely on the collective decisions of governments and space agencies to maintain robust tasking capacity, fund archiving and rapid-dissemination systems, and train local responders who translate satellite products into field decisions. The technology works. The question is always whether the systems, funding, and human capacity around it are adequate to the need.

Energy, Utilities, Methane, and Shipping

Energy infrastructure is one of the largest physical footprints on Earth. Oil and gas pipelines cross tens of thousands of kilometers of terrain. Transmission line corridors run from power plants to cities. Offshore oil platforms sit in dozens of countries' territorial waters. Liquefied natural gas tankers move between ports on schedules that affect global energy prices. Monitoring all of this from the ground, or even from aircraft, is an enormous undertaking. Satellites have become the practical solution.

Methane monitoring is the application that has grown fastest in the past five years, driven by scientific urgency and regulatory pressure on the oil and gas industry. Methane is a powerful greenhouse gas, approximately 80 times more potent than carbon dioxide over 20 years, and it leaks from natural gas production, processing, and distribution infrastructure at rates that vary enormously from facility to facility. Until recently, measuring those leaks required either ground-based sensors or aircraft, both of which were expensive and spatially limited. Two satellite systems have changed that calculation.

GHGSat (Greenhouse Gas Satellite) is a Canadian company operating a constellation of small satellites carrying imaging spectrometers tuned to the absorption wavelengths of methane, carbon dioxide, and water vapor. At its finest resolution, a GHGSat satellite can detect and quantify a methane plume from a single compressor station or landfill, locating the source to within a specific piece of equipment. GHGSat sells monitoring services to oil and gas companies, landfill operators, and regulators who want independent verification of facility emissions.

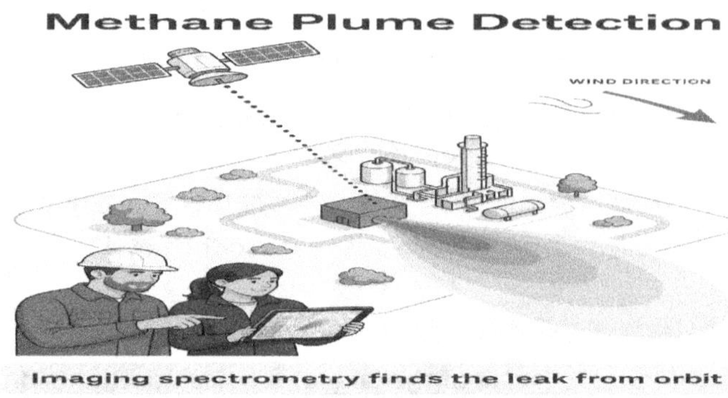

Diagram 7.10: Methane Plume Detection

MethaneSAT is a satellite mission funded in part by the Environmental Defense Fund and launched in March 2024. Where GHGSat specializes in point-source detection at fine resolution, MethaneSAT is optimized for wide-area methane mapping across large oil and gas basins. It surveys an entire production region in a single pass, producing basin-wide emission totals that can be compared to self-reported industry figures. The combination of basin-scale inventory from MethaneSAT and point-source attribution from GHGSat gives regulators and researchers complementary tools: one tells you how much is leaking across a region, the other tells you which specific facility is responsible. Amara's work that afternoon fit this pattern exactly. Her GHGSat analysis of a German gas distribution network found several systematic discrepancies between satellite detections and operators' self-reported emissions, providing the regulator with the evidence needed to require targeted inspections.

Power transmission management is another major application. Utility companies use high-resolution imagery combined with LiDAR point clouds to manage vegetation encroachment in their transmission-line rights-of-way. Trees and tall shrubs growing close to high-voltage lines pose fire and reliability risks, and manual inspection of thousands of kilometers of corridor at the required frequency is impractical. Automated analysis that flags vegetation height exceeding defined clearance thresholds is now the standard approach at major utilities. Solar and wind energy development rely on satellite-derived terrain models, land-cover classification, solar irradiance data, and wind speed estimates derived from reanalysis datasets to site projects and estimate annual energy production.

Global ship traffic is tracked through a system called AIS, which stands for Automatic Identification System. AIS is a maritime safety protocol requiring commercial vessels above a certain size to broadcast their position, identity, speed, and heading at regular intervals via VHF radio. Satellites with AIS receivers can pick up these broadcasts anywhere in the world, and space-based AIS, operated by several commercial providers as well as the European Space Agency, provides a near-real-time picture of global vessel movements that was previously unavailable.

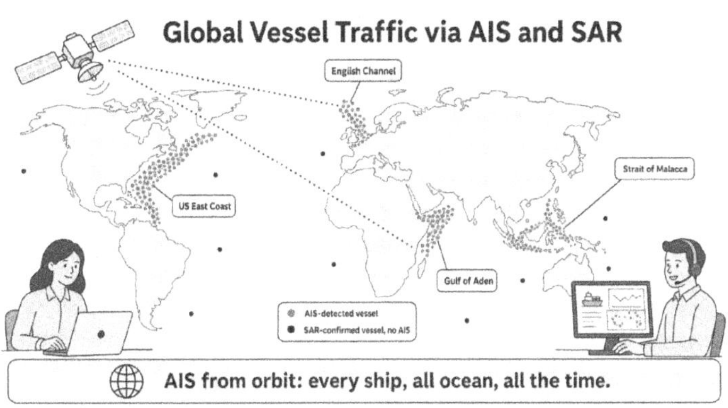

Diagram 7.11: Global Vessel Traffic via AIS and SAR

AIS has limitations. Ships can deliberately turn off their transponder, a practice sometimes associated with sanctions evasion or illegal fishing. SAR steps in as a complementary layer, detecting any vessel large enough to produce a significant radar cross-section regardless of whether it is broadcasting AIS. By comparing SAR-derived vessel detections against AIS position reports for the same time window, analysts can identify "dark" vessels: ships visible in SAR imagery but absent from the AIS record. This technique has been used to monitor sanctions compliance, detect illegal fishing in protected marine areas, and track ship-to-ship cargo transfers, a common method for obscuring the origin of oil cargoes. Pipeline monitoring from satellite data complements both optical and radar time series over pipeline corridors, detecting ground subsidence, vegetation die-off from hydrocarbon contamination, and unauthorized construction near buried lines, all before a failure occurs.

The Pattern Under All Applications

Stand back from the seven sections above and look for what they have in common. You have read about a corn grower's irrigation schedule, a city planner's tree-planting priorities, an ice sheet's mass loss, a hedge fund's parking-lot count, a humanitarian mapper's building traces, and an energy regulator's methane audit. The applications could hardly be more different. And yet they all run on the same assembly of ideas.

Every one of these applications begins with a choice of sensor. Agriculture uses a multispectral sensor and a radar sensor because plant health shows up in chlorophyll reflectance and soil moisture in microwave backscatter. The heat-island application chooses a thermal sensor because the relevant signal is emitted heat, not reflected light. The methane application uses an imaging spectrometer tuned to methane absorption bands because it is the only sensor capable of separating methane from other atmospheric constituents. The sensor choice is dictated by the physics of how the target material interacts with electromagnetic energy.

After the sensor, every application needs a spectral signature or a change pattern that discriminates the thing you care about from everything else. NDVI is a spectral signature for plant health. The diurnal temperature cycle is a pattern of change that distinguishes urban surfaces from vegetated ones. A methane absorption feature at 2.3 micrometers is a spectral signature that no other major atmospheric gas shares. The ship silhouette in a SAR image is a radar scattering signature distinct from the surrounding calm ocean. Without a discriminating signature, the sensor produces data but cannot produce knowledge.

Ground truth is the step that connects satellite observation to physical reality. You cannot trust a model trained only on itself. At some point, someone walks into a field with a handheld spectrometer, a soil probe, and a GPS unit. Someone takes a water sample from a flooded street. Someone inspects a pipeline joint. Someone confirms that the building the AI classified as destroyed is, in fact, destroyed. Ground truth is the tether that keeps the satellite analysis from floating off into numerical fiction.

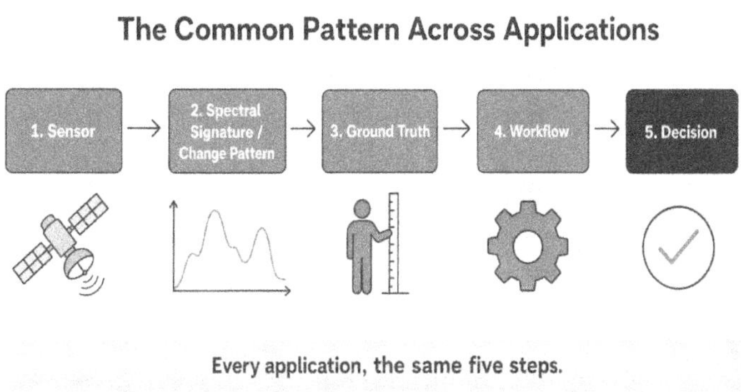

Diagram 7.12: The Common Pattern Across Applications

The workflow is the engineering layer that turns a validated satellite analysis into something a non-specialist can use. It includes the image-processing pipeline, the change-detection algorithm, the classification model, and the final visualization, whether that is an NDVI map on a farmer's phone, a damage-density map on an incident commander's tablet, or a methane plume report in a regulator's inbox. The workflow is where the majority of commercial value is created in the geospatial industry, because raw imagery is becoming cheaper by the year while the ability to deliver actionable insight from that imagery remains scarce.

The decision justifies all the preceding steps. A sensor that does not feed a decision is a hobby. A model with no operational path is a science project. The discipline of connecting satellite observation all the way through to a specific human choice, whether to irrigate, where to plant trees, whether to open a cooling center, which facility to inspect, which vessel to board, is what separates geospatial intelligence from geospatial data. That distinction matters in every context from humanitarian response to commodity trading.

The pattern repeats because the physics does not change. Electromagnetic energy interacts with matter in ways governed by quantum mechanics and thermodynamics, and those interactions have been the same since the first Landsat launched in 1972. What has changed is the density of the sensor network, the resolution of the imagery, the speed of data delivery, the power of the algorithms, and the breadth of people and organizations who can now afford to participate. Once you see the pattern, you can apply it to any new domain you encounter.

Takeaway

This chapter covered a lot of ground because the field does. A technology that began as a Cold War observation tool has permeated the fabric of global agriculture, urban planning, climate science, financial markets, disaster response, and energy regulation. The diffusion has been quiet, mostly invisible to the general public, but it is now deep enough that removing satellite data from any of these industries would cause serious disruption.

Three concepts deserve to stay with you. The first is breadth: geospatial intelligence is not a niche capability for specialized scientists and defense contractors. It is infrastructure, as foundational to modern decision-making as weather forecasts and financial data. The second is unity: beneath the breadth, the same five-step pattern (sensor, signature, ground truth, workflow, decision) repeats in every domain. Learning to see that pattern means you can walk into any geospatial context, from precision agriculture to methane auditing, and quickly understand how the pieces fit together. The third is access: the barriers to using satellite data have fallen dramatically in the last decade. Free imagery archives, commercial data APIs, and cloud processing platforms have democratized access to a degree unimaginable in 2010.

Chapter 8 will take you inside the organizations that produce and sell this intelligence, from national space agencies to the new generation of commercial satellite operators, and will give you the vocabulary you need to evaluate them critically. You now know what the technology can do. Next comes the question of who controls it, who pays for it, and what the rules are.

8 Eyes in the Sky for National Security

Scenario

Theo Lindgren keeps a folder on his laptop labeled "Feb 24, first 48." He rarely opens it at work, but when Marcus and Priya ask him, after a long Thursday debrief, what geospatial intelligence actually looks like in action, he opens it without ceremony. The images are not classified. They never were. That is, he tells them, exactly the point.

The first frame is a synthetic-aperture radar (SAR) image, collected in the early-morning dark of February 24, 2022. SAR is an active sensor that broadcasts its own microwave signal and reads the reflected signal, so clouds and night mean nothing to it. In the image, a column of vehicles stretches for kilometers along a highway north of Kyiv. The shapes are bright against the dark ground, corner reflectors bouncing radar energy straight back at the satellite like a row of mirrors. Capella Space had published this image commercially within hours of collection. Anyone with an internet connection could download it. Theo zooms in. He does not need to say what the vehicles are.

The second frame is a thermal infrared image, captured by a different sensor the same morning, showing intense heat blooms at multiple points within Ukrainian territory. Artillery positions, Theo explains, leave a distinctive infrared signature when tubes cycle through rapid fire. The third frame is a high-resolution optical image of a staging area taken days earlier, showing supply caches, field hospitals, and armored vehicles arranged in the dispersal pattern of a military force preparing to move. Commercial imagery companies had collected it, posted it, and journalists had amplified it globally before any government press office said a word. Marcus leans in toward the screen. Priya says, quietly, that she had seen those images on her phone. She had not known what she was looking at.

Why It Matters

Most people who follow the news have now seen satellite images of conflict zones, disaster sites, or disputed facilities. They arrived in news feeds as screenshots and Twitter threads, labeled with coordinates and arrows. They looked authoritative. Many of them were. But most readers have never been told where those images came from, who collected them, what the images could and could not show, or what rules, if any, governed their release. This chapter answers those questions.

Geospatial intelligence (GEOINT, pronounced "jee-oh-int") is no longer a niche defense specialty. It sits at the center of how the world monitors itself: how arms-control inspectors verify treaty compliance, how journalists document mass atrocities, how environmental watchdogs track illegal deforestation, how governments plan military operations, and how ordinary citizens form opinions about events they will never witness in person. A reader who understands GEOINT can evaluate what a satellite image proves and what it does not prove, recognize responsible OSINT (open-source intelligence) analysis from irresponsible speculation, and hold vendors and governments accountable for how they use these tools. That is the goal of this chapter.

From CORONA to Today: A Brief History

The story of geospatial intelligence begins not with a satellite but with a bucket of film dropping from orbit.

In August 1960, a U.S. Air Force recovery capsule splashed down in the Pacific Ocean carrying a canister of photographic film. The film had been exposed over the Soviet Union by a camera aboard a satellite called CORONA, a program so secret that its existence was not officially acknowledged until 1995. That single mission returned more photographic coverage of the Soviet Union than all U-2 reconnaissance flights combined. CORONA changed everything. For the first time, policymakers could look at denied territory without risking a pilot, without violating airspace in a way that could be publicly denied, and at a scale and consistency that tactical aircraft could never match.

CORONA ran from 1959 to 1972, collecting imagery from more than 100 missions. The program pioneered the basic architecture that still underlies satellite reconnaissance: a camera in orbit, a film- or data-return mechanism, ground processing, and an intelligence consumer awaiting the product. The imagery revealed Soviet missile construction, military bases, ship movements, and industrial facilities. It gave American presidents something concrete to argue from during the Cuban Missile Crisis. It also quietly underwrote arms-control negotiations because both sides knew that cheating on a treaty could now be detected from space.

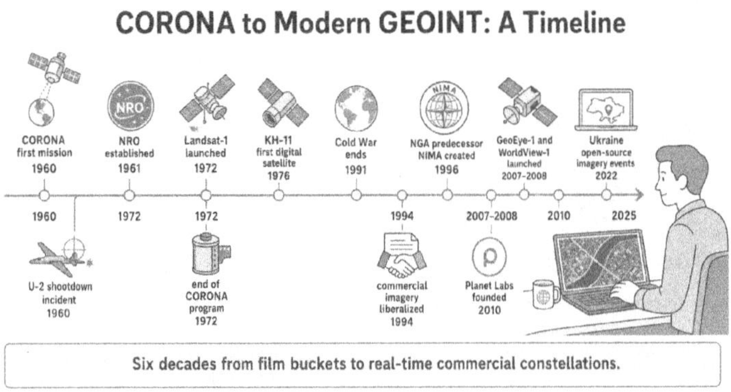

Diagram 8.1: CORONA to Modern GEOINT: A Timeline

CORONA's film bucket approach was elegant but slow. A satellite would collect imagery on one orbit and could only deliver it when the recovery capsule was retrieved from the ocean, a process that took days. The next leap came with KH-11, the first U.S. reconnaissance satellite to transmit digital imagery directly to the ground in near-real time. KH-11 flew its first mission in 1976 and remained classified for many years. Its basic architecture, a large telescope in low orbit relaying digital data to ground stations, remains the foundation of the most capable government reconnaissance satellites today.

Throughout the 1960s and 1970s, the Soviet Union ran its own parallel film-return reconnaissance program, the Zenit series. France began flying the SPOT commercial satellite in 1986. Israel launched its Ofeq series in 1988. India's IRS system also provided civilian imagery beginning in 1988. Reconnaissance from space was, by the end of the Cold War, something many nations understood how to do, even if only the United States and the Soviet Union had the most capable systems.

The Cold War's end reshaped the landscape. The 1994 Presidential Decision Directive PDD-23 authorized U.S. companies to sell commercial satellite imagery at resolutions previously reserved for government systems, starting at 1 meter per pixel. That policy shift launched the commercial Earth observation industry. Companies could now build satellites, sell imagery to anyone, and operate as businesses rather than government contractors. The first major commercial high-resolution satellite, IKONOS, launched in 1999. GeoEye-1 and the WorldView series followed in the mid-2000s. Planet Labs, founded in 2010 and flying the first CubeSat-based imaging constellation, changed the economics again by prioritizing daily revisit over resolution. The history of GEOINT is, in miniature, the history of a technology born in secrecy, matured in government programs, and then partially released into the commercial world with consequences nobody fully anticipated.

NGA, NRO, and How U.S. GEOINT Works

Two agencies sit at the heart of U.S. geospatial intelligence, and most people confuse them or conflate them into a single vague notion of "spy satellites." Understanding their distinct roles makes the whole system clearer.

The National Reconnaissance Office (NRO, pronounced as three letters) designs, builds, launches, and operates the United States' reconnaissance satellites. The NRO was itself a classified organization from its founding in 1961 until its existence was publicly acknowledged in 1992. Its budget and specific capabilities remain classified. What is publicly known is the framework: the NRO operates a fleet of imaging, signals intelligence, and ocean surveillance satellites. It is responsible for the hardware in orbit and for getting the data from space to the ground. Think of the NRO as the engineering and operations shop. It builds the telescope and keeps it flying.

The National Geospatial-Intelligence Agency (NGA) is the intelligence analysis shop. NGA takes data from NRO satellites, commercial satellites, airborne sensors, mapping databases, and open sources, and turns that raw material into finished intelligence products for decision-makers: battlefield commanders, policymakers, emergency managers, and diplomats. NGA also produces the foundational geospatial data that the U.S. military relies on for navigation, targeting, and logistics. Every GPS-guided munition that hits its target is hitting coordinates that NGA's predecessor agencies mapped and verified. NGA employs thousands of imagery analysts, cartographers, data scientists, and geospatial engineers. It is one of the larger members of the U.S. intelligence community by workforce size.

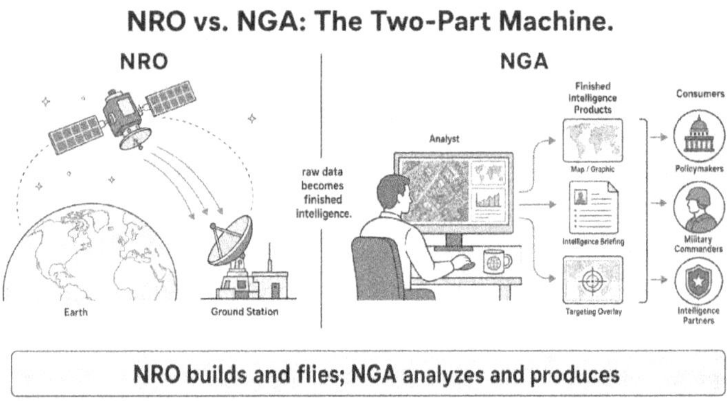

Diagram 8.2: NRO vs. NGA: The Two-Part Machine

The formal discipline of GEOINT is defined in U.S. law as "the exploitation and analysis of imagery and geospatial information to describe, assess, and visually depict physical features and geographically referenced activities on the Earth." That definition encompasses much more than satellite photos. GEOINT products can include: geospatial databases (the digital representation of the Earth's surface), imagery intelligence (IMINT, pronounced "im-int"), mapping and charting products, environmental assessments, and near-real-time imagery of dynamic situations. GEOINT analysts are trained to read imagery the way a cardiologist reads an EKG: they know what normal looks like, and they notice when something deviates from baseline.

A common misconception is that NGA and the intelligence community have exclusive access to the world's best satellite imagery. That was true until roughly 2010. It is no longer true. Maxar Technologies operates the WorldView constellation at 30-centimeter resolution. Planet Labs operates more than 200 imaging satellites covering the entire Earth daily. Capella Space and ICEYE operate commercial SAR constellations that image in any weather, day or night. BlackSky operates a small-satellite optical constellation emphasizing rapid revisit. All of these companies sell imagery to the U.S. government and to any qualified commercial customer. A defense contractor, a hedge fund, a journalist, a humanitarian organization, and a university researcher can all be customers of the same commercial satellite on the same day. The intelligence community's advantage today is not exclusive access to the sensors. It is the classified analytical tools, the all-source fusion capability (meaning the ability to combine GEOINT with signals intelligence, human intelligence, and other sources), and the classified satellite systems whose resolution and capabilities go beyond what any commercial system has publicly disclosed.

NGA's relationship with commercial providers was formalized through a series of contracts. The EnhancedView program, launched around 2010, gave NGA license rights to vast amounts of commercial imagery from DigitalGlobe (now Maxar). The EOCL (the Earth Observation Commercial Landscape) contracts that followed broadened the government's access to the full range of commercial providers. These agreements are significant because they mean U.S. government analysts are working with the same images that commercial and open-source analysts can access. The analytical product is different; the raw material, at least for unclassified work, is increasingly shared.

The Commercial Revolution and Dual-Use Providers

The commercial Earth observation industry did not emerge because governments decided to share their toys. It emerged because the underlying technology became affordable, launch costs dropped with SpaceX and other providers, and a generation of engineers and entrepreneurs realized that persistent, affordable, high-quality Earth observation was a business, not just a government program.

Maxar Technologies, headquartered in Colorado, operates the highest-resolution commercial imaging satellites publicly available. The WorldView-3 satellite, launched in 2014, can resolve ground objects as small as 31 centimeters per pixel in panchromatic (black-and-white) mode. At that resolution, you can identify vehicle types, count individual aircraft on an airfield, distinguish agricultural field boundaries, and monitor construction progress in detail that would have been classified only ten years earlier. Maxar sells imagery to government intelligence agencies, commercial real estate firms, precision agriculture services, mapping companies, and news organizations. The same image of an airfield can appear in a defense intelligence briefing on Monday and in a newspaper story on Wednesday.

Planet Labs operates a very different business model. Planet's "Dove" CubeSat constellation, which grew from a few satellites to more than 200 by the early 2020s, images the entire Earth's landmass every day at approximately three to five meters per pixel resolution. Three to five meters is not enough to identify individual people, but it is enough to see a ship in a harbor, a new road cut through a forest, the extent of a flood, or a military convoy. Planet's core value proposition is frequency, not resolution. By comparing yesterday's image to today's, analysts can detect change anywhere on the planet within 24 hours. Planet sells to agricultural companies tracking crop health, to financial analysts estimating retail traffic from parking lot counts, to environmental NGOs monitoring deforestation, to humanitarian organizations tracking displacement, and to government agencies monitoring military activity.

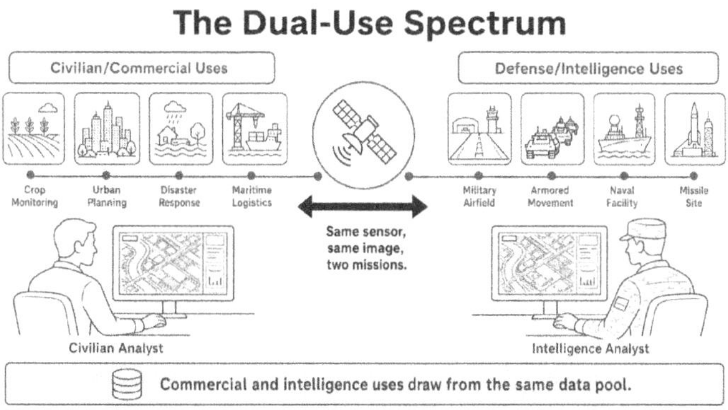

Diagram 8.3: The Dual-Use Spectrum

Capella Space and ICEYE are the two most prominent commercial SAR providers. SAR, or synthetic-aperture radar (covered in detail in Chapter 3), transmits microwave energy toward the ground and records the reflected signal. Because SAR generates its own illumination rather than relying on sunlight, it works equally well at 2 a.m. or 2 p.m., under cloud cover, through smoke and haze. In a conflict or disaster scenario, where operators need continuous monitoring and weather and time of day are unreliable, SAR is invaluable. ICEYE, a Finnish company, can image any point on Earth within a few hours with its SAR constellation. Capella Space, based in San Francisco, offers fine-resolution SAR imagery at centimeter-level precision in its highest-quality mode. Both companies serve government and commercial customers. The same SAR imagery used to track illegal fishing vessels in the Pacific can be used to track military convoys in Eastern Europe.

BlackSky operates a rapid-revisit small-satellite constellation that enables tasking a satellite and receiving an image within 90 minutes of request. This near-real-time responsiveness is particularly valuable in dynamic situations: an analyst who needs to know what is happening at a specific location right now, not six hours ago, can request a BlackSky task and receive an image faster than with most alternatives. BlackSky customers include the U.S. Department of Defense, emergency management agencies, and commercial intelligence services.

The dual-use nature of commercial GEOINT is not a flaw in the system. It is a feature, or at least an inherent property. The economics of satellite operation require broad customer bases. A constellation funded entirely by intelligence agency contracts would likely not exist at the scale and capability that the commercial market now sustains. The commercial ecosystem, paradoxically, produces better intelligence products for government customers than a purely government-funded approach would have. The cost is that the same imagery and capability are available to any customer worldwide who can afford them, including adversarial governments, criminal organizations, and irresponsible actors. Managing that tension is the job of policy frameworks, export controls, and industry governance structures.

Open-Source Intelligence and Citizen Analysts

One of the defining features of the post-2014 GEOINT landscape is the rise of open-source intelligence analysts who work entirely with publicly available data and publish their findings to a global audience in real time. This is a genuinely new phenomenon, and it has changed how conflicts are documented, how governments are held accountable, and how ordinary people experience events happening thousands of kilometers away.

Open-source intelligence (OSINT, pronounced "oh-sint") is intelligence derived entirely from publicly available sources: satellite imagery, social media posts, flight-tracking databases, ship-tracking data, commercially available maps, photographs uploaded by participants or bystanders, and archived news and government records. OSINT is not new. Intelligence agencies have always read foreign newspapers and monitored public broadcasts. What changed is the quantity and accessibility of public geospatial data, the sophistication of freely available analytical tools, and the reach of social media platforms that allow OSINT findings to reach millions of people within hours.

Bellingcat, founded in 2014 by Eliot Higgins, became the most prominent example of civilian OSINT analysis. Higgins began by using freely available satellite imagery and social media photographs to geolocate events in the Syrian civil war, verifying claims by armed groups and governments about where specific attacks took place and what weapons were used. Bellingcat's methodology, which they publish openly, involves cross-referencing imagery from multiple sources, matching geographical features visible in photographs to satellite imagery and map data, and building a documented evidentiary record that any reader can check. Bellingcat investigators used these techniques to identify the specific Buk missile launcher that shot down Malaysia Airlines Flight MH17 over Ukraine in 2014, connecting it to a specific Russian military unit. They identified the officers involved in the 2018 Salisbury poisoning in the United Kingdom using passport records, hotel logs, and publicly available photographs. These investigations, conducted by journalists with no access to classified information, reached conclusions that governments with full intelligence resources had reached independently but could not publicly disclose.

Diagram 8.4: The OSINT Workflow

The @OSINTtechnical account on X (formerly Twitter) represents another type of practitioner: a highly skilled individual analyst who posts real-time imagery analysis during developing events. During the Russian invasion of Ukraine beginning in February 2022, accounts like @OSINTtechnical, @noclador, and many others were routinely posting geolocated imagery analysis faster than traditional news organizations could verify and publish comparable information. These analysts used commercial satellite imagery (much of it released in near real time by companies such as Maxar and Planet), Sentinel-1 SAR imagery from the European Space Agency's Copernicus program, and photographs posted by military personnel and civilians on social media. They matched landmarks, road features, building patterns, and geographic terrain features to identify exactly where specific events were occurring, which military units were involved, and, in some cases, what equipment was being used.

The conflict in Ukraine became the first large-scale war in which commercial satellite imagery and OSINT analysis played a major public role. Images published by Maxar and Planet of the Bucha area north of Kyiv, showing bodies in the streets of a suburb that Russian forces had recently occupied, became key evidence in international discussions about war crimes. The imagery was publicly available, covered by major news organizations, and analyzed simultaneously by independent researchers and government experts. The evidentiary standard it represented was a departure from previous conflicts, where visual documentation was often contested or inaccessible.

The OSINT community is not uniformly skilled or responsible. The same accessibility that allows a skilled analyst to geolocate a verified event also allows an amateur or bad-faith actor to post convincing-looking but incorrect analysis. Misidentified locations, mislabeled imagery from previous conflicts, and weaponized disinformation using satellite imagery have all appeared in major conflicts. Evaluating OSINT requires the same critical habits as evaluating any source: Who is the analyst? What is their track record? What are the primary sources? Is the methodology documented so that others can check the work? Can independent analysis corroborate the findings? These questions matter more, not less, when the imagery looks authoritative.

The citizen-analyst community has also contributed to arms-control monitoring. The Arms Control Wonk community and researchers at the James Martin Center for Nonproliferation Studies have published satellite imagery analysis of nuclear facilities, missile test sites, and military installations in North Korea, Iran, China, and elsewhere. In several instances, independent researchers published findings about new missile silo construction or nuclear facility expansion before any government had made a public statement. The imagery was commercially available to anyone. The expertise to interpret it was learned, shared, and applied by researchers outside the intelligence community.

The Technical Mix in Modern GEOINT

A GEOINT product assembled for a real-world decision-maker is rarely based on a single sensor or a single image. Modern GEOINT draws on a layered technical stack, and understanding the components helps a reader evaluate what any given intelligence product can and cannot tell you.

Very high resolution (VHR) optical imagery sits at the top of the resolution hierarchy in the commercial world. Satellites like WorldView-3 and WorldView-Legion can image at 30 to 50 centimeters per pixel under clear conditions. At that resolution, an analyst can identify aircraft types on an airfield, measure the dimensions of a construction project, distinguish between military vehicle types, and, in many cases, read container labels on a dock. VHR optical imagery is available commercially and through government contracts. Its limitation is the physics of passive optical sensors: it requires daylight and clear skies. Cloud cover, smoke, haze, or darkness renders an optical image unusable or degraded.

SAR fills the gaps optical cannot cover. As covered in Chapter 3, SAR transmits its own signal and measures the return, so clouds and darkness are irrelevant. SAR imagery looks different from optical imagery: buildings, vehicles, and other corner reflectors appear as bright spots; water appears dark; vegetation and soil have characteristic textures based on their roughness and moisture content. A trained SAR analyst reads these signatures the way a dermatologist reads skin texture: the patterns have meaning, and deviations from pattern matter. Commercial SAR at 50-centimeter resolution or better can detect individual vehicles and vessels, identify ship types, monitor industrial facility activity, and track movement in denied areas. The Ukraine scenario in this chapter's opening scene used SAR imagery precisely because the invasion began at night, in winter, under cloud cover. Optical imagery would have shown nothing; SAR would have shown everything.

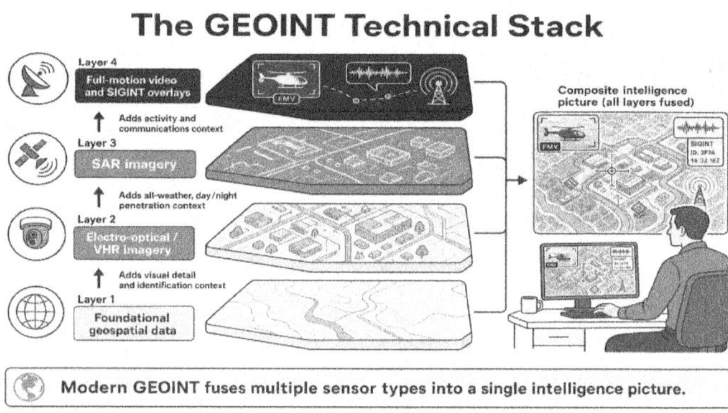

Diagram 8.5: The GEOINT Technical Stack

Full-motion video (FMV) is the third major sensor type in the GEOINT stack. FMV refers to continuous video footage of a geographic area, originally collected primarily by drone-mounted cameras but increasingly available from satellite platforms. Unlike a still image, FMV shows movement over time: vehicles entering and leaving a facility, people gathering, aircraft taxiing, ships maneuvering. FMV is particularly valuable for pattern-of-life analysis, where analysts watch a location repeatedly over days or weeks to understand the routine of what happens there, so that a deviation from routine can be recognized as significant. Military drone operations in conflict zones rely heavily on FMV. Commercial FMV from satellites is an emerging capability; several companies are developing video-capable small satellites that can record short clips of specific locations on request.

Signals intelligence (SIGINT) provides a layer that pure imagery cannot: information about electronic emissions. SIGINT includes communications intelligence (intercepted messages), electronic intelligence (radar emissions, navigation signals), and measurement and signature intelligence. In a GEOINT context, SIGINT overlays on imagery can tell an analyst not just that vehicles are present at a location but that certain radio frequencies are active, certain radar systems are operating, or certain communication patterns are occurring. SIGINT in the U.S. system is the primary domain of the National Security Agency (NSA), but fusing SIGINT indicators with GEOINT products is a core all-source intelligence function. Commercially, some companies offer automatic identification system (AIS) data, the maritime transponder signal that ships broadcast for navigation safety, overlaid on SAR imagery. AIS, combined with SAR, allows an analyst to see all vessels in an area and identify which ones are broadcasting their location versus which are "dark" and attempting to avoid detection. A fishing vessel that turns off its AIS transponder in an area of known illegal fishing is behaving anomalously. A naval vessel that goes dark before entering a disputed area is doing so deliberately. The combination of imagery and signals data makes both situations visible.

Automatic Identification System (AIS) data warrant elaboration because it illustrates the fusion principle well. AIS was designed as a maritime safety protocol: ships transmit their identity, position, speed, and heading continuously so that other ships and port authorities can track them. AIS data is publicly available through services like MarineTraffic and VesselFinder, which aggregate the signals into global ship tracking maps accessible to anyone. When an analyst overlays AIS data on SAR imagery, they get two independent datasets about the same ships at the same moment. A ship that appears on the SAR image but not in the AIS data is a "dark vessel," one that has turned off its transponder or never had one. Dark vessels are disproportionately associated with sanctions violations, illegal fishing, and military operations. This is civilian data used for intelligence purposes. The ship owners did not intend to advertise their presence to analysts, but the physics of radar made them visible regardless of what they chose to transmit.

Law, Ethics, and the Dual-Use Dilemma

The capabilities described in this chapter raise legitimate questions that deserve honest answers. Who decides what can be imaged and sold? What prevents commercial GEOINT from enabling atrocities? Are there rules? Do they work?

The short answer is: yes, there are rules; they are imperfect; they are being revised constantly; and the people responsible for them know the challenges are outpacing the governance frameworks. That is not a comfortable answer, but it is the accurate one.

In the United States, commercial satellite imagery is regulated by the National Oceanic and Atmospheric Administration (NOAA), which licenses commercial remote sensing operators under the Land Remote Sensing Policy Act and subsequent regulations. Operators must obtain a license, may not image in ways that endanger national security, and may be subject to "shutter control," a provision that allows the government to restrict imaging of specific areas during national security emergencies. Shutter control was invoked briefly during Operation Desert Storm in 1991 and has been debated but rarely used since, largely because the globalization of commercial imagery has made unilateral U.S. shutter control less effective: a foreign commercial provider might image the same area even if a U.S. provider is restricted.

The Kyl-Bingaman Amendment, passed in 1997 and updated in 2020, restricted U.S. commercial satellite operators from selling imagery of Israel and the Occupied Palestinian Territories at resolutions better than those available from comparable commercial providers globally. The original restriction set a floor of two meters per pixel, intended to prevent U.S. commercial operators from providing adversaries with intelligence-quality imagery of Israeli territory at a time when the global commercial market did not yet offer better imagery. The 2020 revision lowered the floor to match current commercial market norms. The Kyl-

Bingaman Amendment illustrates the difficulty of export control in a global market: it applies only to U.S. operators, and as European, Asian, and other commercial providers reached similar resolutions, the U.S.-specific restriction became commercially disadvantageous without providing commensurate security benefit.

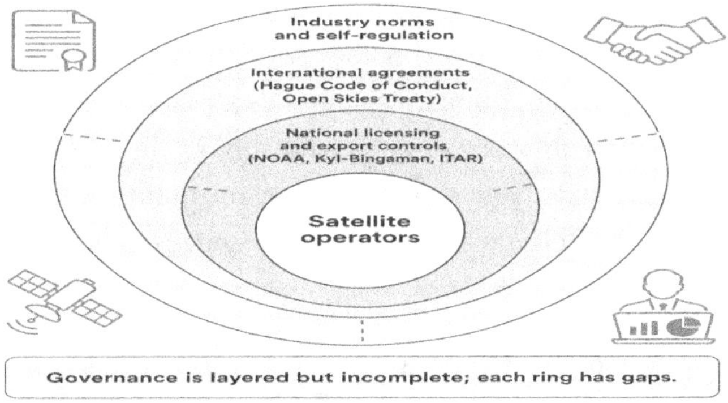

Diagram 8.6: The Legal and Governance Landscape

The Hague Code of Conduct against Ballistic Missile Proliferation (HCoC, subscribed to by more than 140 countries) is primarily a transparency instrument for missile programs. Still, it reflects the broader principle that states should provide notification of space launches and maintain confidence-building measures around space-based military activities. It is not a treaty in the formal sense and has no enforcement mechanism, but it establishes norms that influence state behavior in the domain where reconnaissance satellites operate.

The Open Skies Treaty, signed in 1992, allowed signatory nations to fly unarmed observation aircraft over each other's territory on prescheduled terms. The treaty was intended to reduce the risk of surprise attacks and build mutual transparency between NATO and former Warsaw Pact nations. The United States withdrew from the treaty in 2020, citing Russian violations. Russia withdrew in 2021. The treaty is effectively defunct, and its demise illustrates what happens when a confidence-building mechanism loses political support: the transparency it provided disappears, and both sides revert to unilateral intelligence collection, commercial imagery, and OSINT to fill the gap.

The International Traffic in Arms Regulations (ITAR) restrict the export of defense-related technology and information, including certain satellite sensor technologies and some forms of processed imagery. ITAR applies to U.S. companies and imposes licensing requirements on sharing sensitive geospatial technology with foreign customers, partners, or employees. Companies that design satellite sensors for government reconnaissance programs must navigate ITAR restrictions on the commercial derivative products they might offer. The regulatory environment is complex and layered, and it evolves as the technology evolves.

Beyond regulation, the harder question is ethics, and it is a question the GEOINT industry discusses with varying degrees of comfort. The dual-use dilemma is real: there is no technical distinction between an image used to coordinate humanitarian aid delivery and an image used to plan a strike. The same Maxar image of a refugee camp, seen by an NGO logistics planner, is, in theory, available to any party in any conflict. The same SAR image that shows a city's flood extent for emergency management also reveals its road network and infrastructure for any adversary planning an attack. Commercial companies have adopted policies restricting access to imagery during active conflicts in some circumstances. Maxar and Planet both implemented policies during the Ukraine conflict affecting the availability of certain imagery to certain customers. Industry-developed norms include not selling imagery that could be used to directly facilitate human rights abuses, but defining and enforcing such norms is difficult when imagery is sold through resellers, data brokers, and application platforms that the original company may not monitor.

The question of persistent surveillance is among the most contested ethical debates in commercial GEOINT. A constellation that images a specific city every few hours could, in theory, be used to track individuals. At a three- to five-meter resolution, individuals cannot be identified. At 30-centimeter resolution, individuals can sometimes be distinguished in a crowd by clothing, context, and movement patterns, though reliable identification is not practically achievable at current commercial resolutions. As satellite constellations grow denser and as video capabilities improve, the theoretical privacy implications become more acute. Regulatory frameworks designed in the 1990s were not designed with near-continuous video coverage in mind.

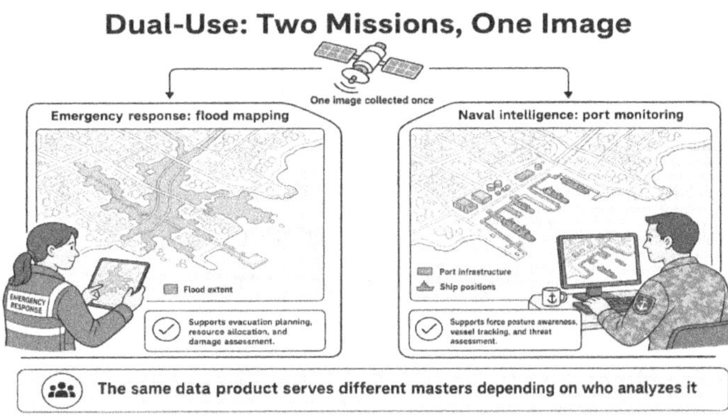

Diagram 8.7: Dual-Use: Two Missions, One Image

The resolution limits built into current commercial licensing reflect a deliberate policy choice to prevent the commercial market from reaching capabilities historically reserved for the most capable government systems. The specific capability thresholds for the most sensitive government reconnaissance satellites remain classified. Analysts believe the gap between the best publicly available commercial imagery and the best government systems is substantial. However, the exact margin cannot be stated with certainty from open sources.

What Informed Citizens Should Know

Geospatial intelligence used to be invisible to ordinary citizens. It happened in classified facilities, appeared in presidential briefings, and shaped events whose public explanations offered no hint of the orbital observations beneath the surface. That invisibility is largely gone. GEOINT is now a daily feature of news reporting, social media, academic research, and public policy debate. This is mostly good. It is also complicated.

Here is what a thoughtful, informed citizen should understand about GEOINT as it relates to public life.

Commercial satellite imagery is powerful but not omniscient. The resolution that commercial providers offer can show vehicle types, building states, and vegetation conditions in remarkable detail. It cannot reliably read license plates, identify specific individuals by face, or resolve objects smaller than 30 centimeters with currently available commercial systems. When news reports describe what satellite imagery shows, a discerning reader should ask: which sensor collected it, at what resolution, and under what conditions? A 3-meter Planet image and a 30-centimeter Maxar image contain very different levels of information even though they may appear visually similar as thumbnails in a news article.

OSINT analysis varies enormously in quality. The best practitioners, organizations like Bellingcat and well-credentialed individual analysts with documented track records, apply rigorous methodology, document their sources, and invite peer review. Their work is often more reliable than first-pass news reporting precisely because it shows the evidence. The worst practitioners post misidentified imagery, draw unsupported conclusions, and sometimes spread deliberate disinformation using authentic-looking satellite images from unrelated events or locations. Before amplifying a piece of OSINT analysis, a careful reader should ask: Is the analyst named and accountable? Are the primary source images available for independent review? Have other independent analysts corroborated the analysis? Does the methodology make sense?

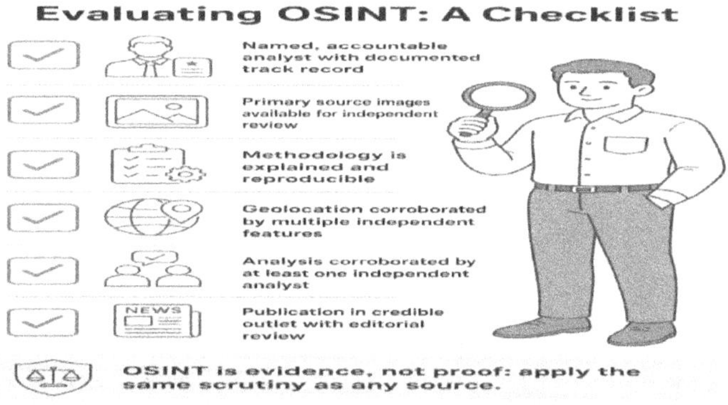

Diagram 8.8: Evaluating OSINT: A Checklist

The commercial providers themselves are accountable actors with business interests, not neutral parties. Commercial satellite companies make decisions about what to image, when to release imagery publicly, what to sell and to whom, and how to respond to government requests to restrict or accelerate access. Understanding that Maxar, Planet, and others operate as businesses with stakeholders, government contracts, and reputational interests helps a reader evaluate their choices. When a company releases dramatic imagery of a conflict area to the press, there is usually a reason beyond altruism, whether that is demonstrating capability to government customers, deterring bad actors through visibility, or serving a news-cycle business interest. None of that makes the imagery less valuable or less accurate, but it contextualizes how and why it becomes available.

Citizens in democratic societies have legitimate interests in how their governments use GEOINT capabilities. Domestic surveillance using satellite imagery is legally restricted in most democracies but is a live policy concern as satellite density increases. Export controls on satellite imagery technology are matters of foreign policy and national security, with consequences for trade, allied relationships, and arms control. The conduct of intelligence operations, including GEOINT operations, is subject to oversight by legislative bodies in most democratic systems. Being an informed citizen in this domain means understanding enough about the technology and the policy framework to evaluate what governments and companies claim, to support good governance, and to push back against uses of GEOINT that are disproportionate, unaccountable, or dishonest.

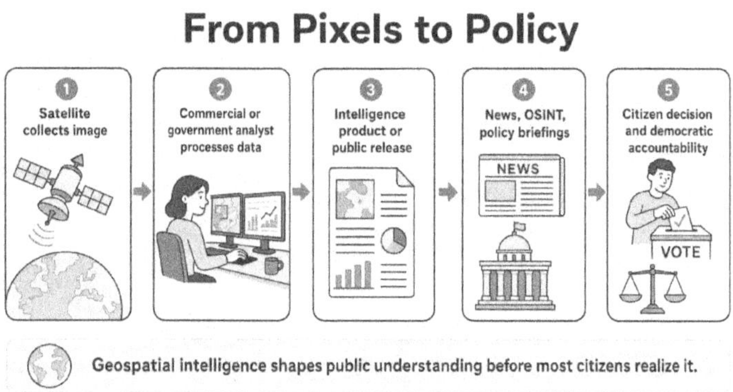

Diagram 8.9: From Pixels to Policy

There are trustworthy resources for citizens who want to follow GEOINT developments without specialized training. The NGA's public website publishes unclassified reports, educational resources, and its GEOINT basic doctrine. The Center for Strategic and International Studies (CSIS) publishes analysis on commercial space and GEOINT policy. The Secure World Foundation tracks space security and satellite governance. Arms Control Wonk publishes satellite imagery analysis of nuclear and missile programs. Bellingcat publishes methodology guides that explain how its investigators work. The Copernicus program, operated by the European Space Agency, provides free access to Sentinel satellite imagery that any member of the public can download and analyze. These resources collectively allow an engaged citizen to move beyond receiving GEOINT analysis secondhand and begin developing their own capacity to evaluate it.

Geospatial literacy is becoming a basic requirement for civic participation, much as media literacy became essential with the rise of digital news. Understanding that a satellite image is not a photograph in the naive sense, that it is a measurement made by a sensor with specific physical properties, collected at a specific time, processed through algorithms that make choices about what to emphasize, analyzed by a person or software system with particular training and interests, and published by an organization with a specific agenda, all of that context determines what the image means. A citizen who understands this is harder to manipulate and better equipped to evaluate the claims made in its name.

Diagram 8.10: The Commercial GEOINT Ecosystem

Takeaway

The chapter you have just read covers a topic that many books avoid because it sits at the intersection of technical complexity, national security sensitivity, and genuine ethical tension. The goal has been to treat you as an informed adult who can handle honest answers.

Geospatial intelligence originated in secrecy, matured in Cold War competition, and has now partially spilled into the commercial world with consequences that governments, companies, and citizens are still working out. The NRO builds satellites and collects data; NGA analyzes it and produces intelligence. Commercial providers like Maxar, Planet, Capella, BlackSky, and ICEYE serve both government and civilian markets with the same hardware. OSINT analysts with laptops and publicly available imagery have changed how conflicts are documented and how accountability operates. The technical stack- VHR optical, SAR,

FMV, AIS, and signals overlays- is more capable and more accessible than it has ever been.

The dual-use nature of this technology is not a solvable problem. It is a permanent feature of a world in which the physics of remote sensing do not distinguish between beneficial and harmful applications. Governance frameworks, export controls, industry norms, and democratic oversight all play roles in managing the dual-use reality, and none is sufficient on its own. The appropriate response for a citizen is not anxiety or cynicism. It is the kind of informed engagement this chapter has tried to model: understanding how the tools work, who controls them, what rules apply, and where the accountability gaps are.

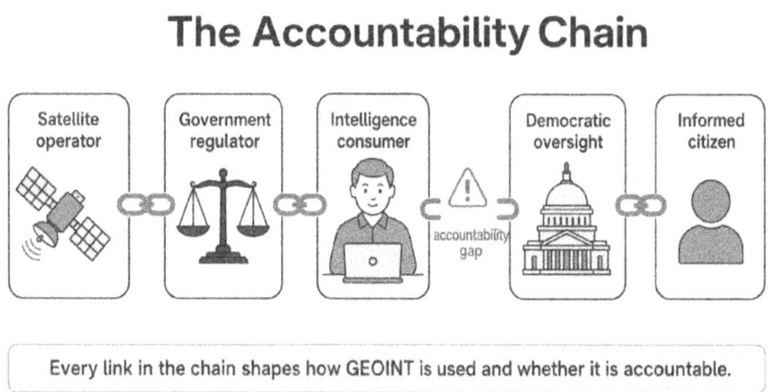

Diagram 8.11: The Accountability Chain

The satellite images that appeared in your news feed during the most recent major crisis did not fall from the sky fully formed. They were collected by sensors operating according to physical laws, processed by algorithms encoding human choices, analyzed by people with training and interests, and released by organizations with business and policy motivations. Every step in that chain is human, fallible, and improvable. Understanding the chain is the beginning of participating in its improvement.

Diagram 8.12: GEOINT for the Informed Citizen

9 Building Your First Geospatial Workflow

Scenario

Marcus Okonkwo's desk was buried in browser tabs. It was a Tuesday afternoon, six weeks into his new role at Pleiades Geospatial Services, and he had just admitted to himself that watching his colleagues talk about spectral bands and vegetation indices was not the same thing as understanding them. He had a background in finance data: clean spreadsheets, well-documented APIs, problems where the domain made intuitive sense. Satellite imagery did not yet make intuitive sense. The files were enormous. The terminology was unfamiliar. And every tutorial he found online seemed to assume he already knew the vocabulary he was trying to learn.

That evening he opened his laptop at home and made a simple decision. He would not try to learn everything at once. He would pick one place he knew well, download one image of it, and ask one question about it. The place was a nature preserve he hiked every spring, twenty minutes outside the city. The question was: does the vegetation here look healthy right now? He found the Copernicus Open Access Hub, typed in the coordinates, and thirty minutes later was staring at a real Sentinel-2 multispectral scene of the woods he had walked dozens of times. The image was free. The software to open it was free. The feeling of finally touching real satellite data was, he would say later, quietly electrifying.

Over the following six weeks, Marcus turned that one image into a small but genuine portfolio. A vegetation-health map. A before-and-after comparison of a construction site that had appeared near the preserve. A short time-series animation of seasonal greenness from the previous three years. None of it required code. None of it required a paid subscription. All of it connected, directly and viscerally, to concepts from the earlier chapters of the books he had been reading. This chapter is his walkthrough.

Why It Matters

Every chapter before this one was preparation for this moment. You have learned how satellites collect light across multiple spectral bands, how indices like NDVI (Normalized Difference Vegetation Index) compress that light into a single health signal, how change detection works, how time series reveal the planet's rhythms. You have the concepts. What you may still be missing is the experience of holding actual data in your hands, opening an actual file, and seeing an actual place rendered in the language of remote sensing. That experience changes something. It closes the gap between knowing and understanding.

The path Marcus followed is reproducible. Every tool he used is free. Every dataset he accessed is public. The skills he built in those sixty days are real analyst skills, the same ones used by people at USGS (the United States Geological Survey), ESA (the European Space Agency), humanitarian NGOs, and city planning departments around the world. This chapter lays out that same path, step by step, so that you can follow it on your own schedule and at your own pace. You do not need a computer science degree. You do not need to write code. You need curiosity, a computer, and a place you care about.

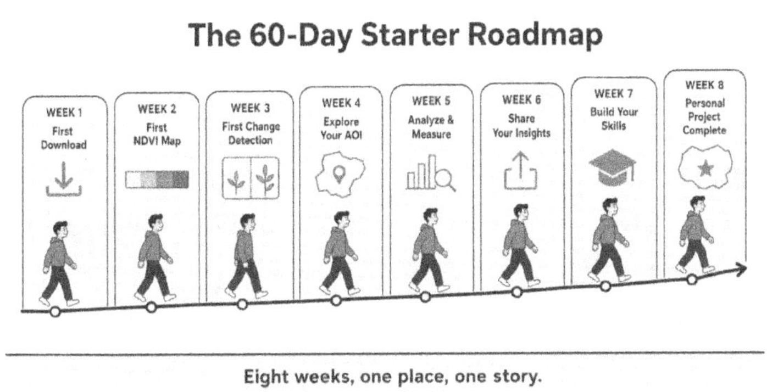

Diagram 9.1: The 60-Day Starter Roadmap

The Free Tools Real Analysts Actually Use

Before you download a single scene, it helps to know the landscape of available tools. The geospatial field has a rich ecosystem of free, professional-grade software, and knowing what each tool does saves you from picking the wrong one for your task. Think of this section as a tool-rack on a workshop wall: you do not need to use every tool on day one, but you should know which one to reach for when a specific job comes up.

QGIS

QGIS (Quantum Geographic Information System, though almost no one uses the full name) is the free, open-source desktop GIS application that most analysts outside large commercial organizations use as their primary tool. It runs on Windows, macOS, and Linux. It opens satellite imagery in dozens of formats, lets you apply color composites and band math, draw and edit vector layers, run basic analysis, and export print-quality maps. The interface takes a few sessions to feel natural, but the learning curve is much gentler than it looks on first launch.

What makes QGIS powerful is its plug-in ecosystem. Hundreds of add-ons extend its core functionality, covering everything from atmospheric correction to classification workflows. The QGIS community is large, welcoming, and well documented. For the exercises in this chapter, QGIS will be your main desktop environment.

Google Earth Engine

Google Earth Engine (GEE) is a cloud-based platform that gives you access to more than fifty years of satellite imagery without downloading anything to your computer. You work through a browser-based code editor or, increasingly, through a no-code interface for simple tasks. The platform holds the entire Landsat archive, the full Sentinel collection, MODIS products, global terrain models, and hundreds of other datasets, and it runs computations on Google's servers rather than on your laptop. This means you can analyze continental-scale datasets in minutes instead of days.

For non-coders, GEE's most accessible entry point is its Explorer interface, which lets you browse imagery and apply basic visualizations without writing a single line of code. Even if you never touch the code editor, the Explorer alone makes GEE useful for quickly assessing what imagery exists for a given place and time period. A free account is available to non-commercial users through a simple registration.

NASA Worldview

NASA Worldview is the fastest way to browse what is being observed from space right now. It is a browser-based viewer with no installation required, and it streams imagery from dozens of instruments, including MODIS (Moderate Resolution Imaging Spectroradiometer), VIIRS (Visible Infrared Imaging Radiometer Suite), and several others, in near real time. You can step backward through dates to review recent events, overlay atmospheric data, sea surface temperatures, wildfire hotspots, and aerosol concentrations, and share a specific view by copying a URL.

Worldview is not designed for precise measurement or detailed analysis, and you cannot download full scenes directly from it. But as a situational-awareness window on the planet, it is unmatched for speed and simplicity. Analysts use it daily to spot active disasters, check cloud cover, and deliver briefings to audiences who need a visual within seconds.

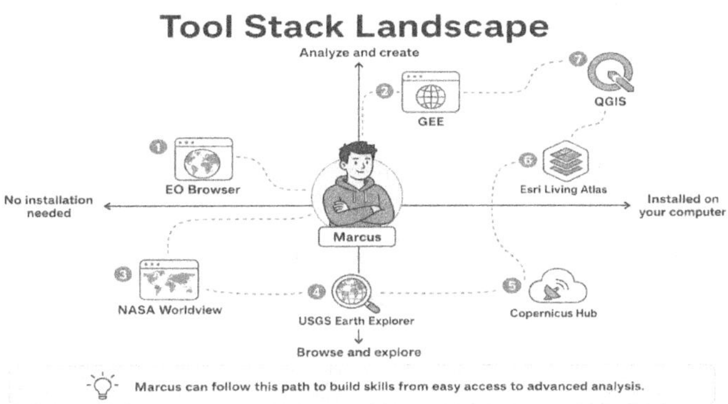

Diagram 9.2: Tool Stack Landscape

USGS Earth Explorer

USGS Earth Explorer is the portal where you download free Landsat scenes, SRTM (Shuttle Radar Topography Mission) elevation data, and a wide range of other US government datasets. Navigation is a bit old-fashioned by modern standards, but the underlying data archive is genuinely extraordinary. You can draw a bounding box on the map, set a date range, filter by cloud cover, and receive a list of every available scene for that combination. Downloads are free and immediate after a free account registration.

The key advantage of Earth Explorer is depth. The Landsat archive goes back to 1972 for some areas, making it the longest continuous record of Earth's land surface from space. For any question that asks what a place looked like decades ago, Earth Explorer is where you start.

Copernicus Open Access Hub

The Copernicus Open Access Hub (sometimes called the Copernicus Data Space Ecosystem after its 2023 relaunch) is the official portal for downloading free Sentinel satellite data from ESA (the European Space Agency). Sentinel-2 imagery, with its 10-meter resolution and 13 spectral bands, is what most beginners should start with. The data are free under an open license, and the hub's interface allows you to search by location, date, and cloud-cover percentage.

Sentinel-2 revisits any point on Earth every five days with two satellites flying in the same orbit. That means for almost any place you want to study, there are hundreds of scenes available. Once you find a scene with acceptable cloud cover over your location of interest, you can download the full product to your computer in minutes. The files are large (a typical Sentinel-2 Level-2A product is several gigabytes) but well structured and easy to open in QGIS.

Sentinel Hub EO Browser

EO Browser, developed by Sentinel Hub (a commercial company with a generous free tier), is the most visually satisfying way to explore Sentinel imagery without downloading anything. You type an address or pan to a location, pick a date, select a visualization preset (true color, false color, NDVI, SWIR composite, and many more), and the imagery renders instantly in your browser. You can compare two dates side by side, animate a time series, and download low-resolution versions of your view for presentations.

EO Browser is where Marcus first saw his hiking preserve in NDVI colors. The tool is immediate and responsive, and it tends to produce those first genuine moments of recognition for beginners. The free tier has daily download limits, but for exploration and learning it is effectively unlimited.

Esri Living Atlas

Esri's Living Atlas of the World is a curated collection of global datasets available through ArcGIS Online, with a portion freely accessible without a paid subscription. The Atlas includes authoritative layers for administrative boundaries, demographic data, real-time weather, ocean temperature, and a growing collection of satellite-derived products. You can explore, visualize, and export many layers through a browser without owning an ArcGIS license.

For beginners, the Living Atlas is most useful as a reference layer. Pull in the official street network, land parcel boundaries, or census demographics to give your satellite-derived analysis the human context that makes the result legible to a general audience.

Screenshot Style: EO Browser Interface

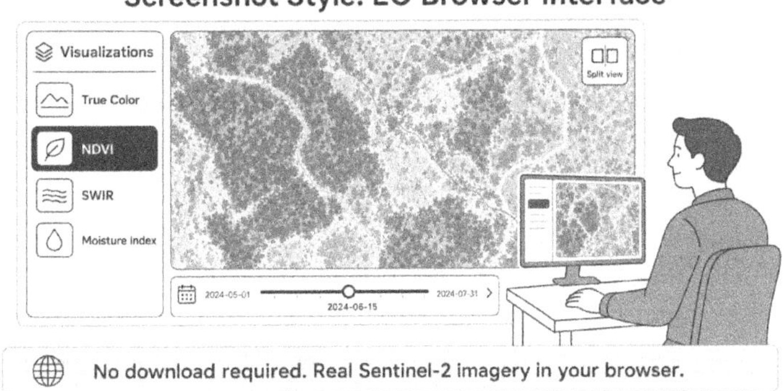

Diagram 9.3: Screenshot Style: EO Browser Interface

Free Training Resources Worth Your Time

Finding good free geospatial training is easy. Finding good free geospatial training that is genuinely useful to a beginner without a technical background is harder. The internet is full of tutorials that assume you already know Python, spend three hours on theoretical background before you ever touch data, or teach a tool version that is now two major releases old. The resources below have been filtered for quality, accessibility, and staying power.

NASA ARSET

NASA ARSET (Applied Remote Sensing Training) is the single highest-value free training resource in the field for most beginners. ARSET runs live webinar-based courses throughout the year and archives every session so you can complete them on your own schedule. The courses cover a wide range of topics: agricultural monitoring, flood mapping, air quality analysis, wildfire detection, ocean color, and urban heat island analysis, among others. Each course consists of three to five sessions, each roughly 90 minutes long.

What makes ARSET stand out is that the instructors are practicing NASA scientists who have designed their courses specifically for professionals without deep technical backgrounds. The target audience includes water resource managers, urban planners, public health officials, and journalists, not software engineers. The courses use real datasets, real tools (often Google Earth Engine or QGIS), and real case studies from ongoing NASA research. Registration is free and open globally. If you take only one structured course from this list, take a NASA ARSET course in your area of interest.

ESA EO Education

ESA's Earth Observation education portal, sometimes called EO College or ESA EO Education depending on the program, offers a growing library of online courses on topics ranging from SAR (Synthetic Aperture Radar) interpretation to land cover classification to glacier monitoring. The courses are MOOC-style (Massive Open Online Course), meaning you progress through videos, readings, and exercises at your own pace. Certificate tracks are available for some programs.

ESA's training goes somewhat deeper into sensor physics than NASA ARSET, which is either an advantage or a challenge depending on your patience for background material. For someone who has worked through the earlier chapters of this book and wants to extend their spectral knowledge, the ESA EO Education portal is a natural next step.

Google Earth Engine Tutorials

The Google Earth Engine documentation includes a full set of interactive tutorials built directly into the GEE code editor environment. Even if you do not intend to write code, the tutorial series for beginners provides an excellent conceptual walkthrough of how GEE organizes satellite data, how image collections work, and how basic operations like band selection and index calculation are structured. Many of the concepts translate directly to what you will do in QGIS, even though the interface is different.

For those willing to dip a toe into GEE's JavaScript editor, the tutorials are well written, start from absolute zero, and produce visible results within minutes. The GEE community forums are active, and a well-described question typically receives a thoughtful answer within a day.

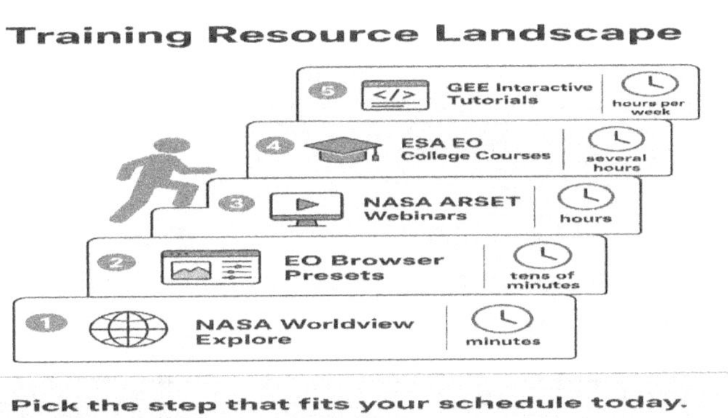

Diagram 9.4: Training Resource Landscape

OpenStreetMap and HOTOSM

OpenStreetMap (OSM) is a free, editable map of the world built by volunteers, often described as the Wikipedia of geographic data. Joining the OSM mapping community is a legitimate form of geospatial training because it forces you to think carefully about spatial accuracy, attribute tagging, and the relationship between what you see in an image and what you record as a data point. The iD editor, OSM's browser-based editing tool, is gentle enough for a complete beginner to contribute within an hour.

HOTOSM (the Humanitarian OpenStreetMap Team) runs regular mapping activations in response to disasters, where volunteers trace buildings, roads, and infrastructure from satellite imagery to help response teams on the ground. The Tasking Manager, HOTOSM's coordination tool, divides large areas into small squares so you can contribute meaningful work in a short session. Many analysts describe their first HOTOSM activation as the moment geospatial work stopped feeling abstract.

YouTube Channels Worth Bookmarking

A curated YouTube approach saves dozens of hours. University remote sensing programs often release full lecture series free of charge. Individual practitioners with consistent posting schedules tend to produce shorter, task-focused videos that answer specific questions efficiently. The QGIS project has an official YouTube channel with tutorial playlists that track the current software version. When evaluating any tutorial, check the publication date: an older video may reference an interface that has since changed significantly.

Week One: Your First Downloaded Scene

Week one has one goal: hold a real satellite scene on your computer. Not a screenshot, not a web map tile, but an actual multispectral data product that you downloaded yourself from a public archive. This is simpler than it sounds, and completing it will change your relationship to every piece of satellite imagery you encounter afterward.

The recommended scene is a Sentinel-2 Level-2A (atmospherically corrected) product over a place you know well. "A place you know well" is important. When you eventually load the image and look at it, you want your brain to engage with it as a real place, not as an anonymous patch of ground. Choose your neighborhood, your hometown, the nearest national park, the coastline you visit in summer, a farm you have driven past. Familiarity multiplies the learning value.

Here are the steps for week one:

1. Go to the Copernicus Data Space Ecosystem (dataspace.copernicus.eu) and create a free account. Registration takes about five minutes and requires only an email address.

2. Once logged in, navigate to the Browser interface. You will see a world map with a search panel on the left.

3. Type your location of interest into the search bar or pan and zoom the map manually until your chosen place is centered.

4. In the search panel, select "Sentinel-2" as your satellite and "L2A" as the processing level. L2A means the atmospheric correction has already been applied, which saves you a step.

5. Set a date range. For your first scene, aim for the summer months in the Northern Hemisphere (June through August) or the dry season if you are in a tropical location. Summer scenes tend to have less cloud cover and show maximum vegetation.

6. Set a maximum cloud cover percentage of 10 to 20 percent. This filters out scenes where clouds would obscure your area of interest.

7. Click Search. A list of available scenes will appear, each with a thumbnail and a cloud cover percentage.

8. Click on a scene with low cloud cover over your location. A footprint will appear on the map showing the scene's coverage area.

9. Click the Download button. The full Sentinel-2 product typically ranges from 600 megabytes to 2 gigabytes, so make sure you have space and a stable connection.

10. While it downloads, install QGIS from qgis.org if you have not already. The download and installation take about fifteen minutes.

When the download finishes, open QGIS, drag one of the band files from the download folder onto the map canvas, and see what you have. It may appear gray or washed out at first. That is normal. You are looking at raw digital number values in a single band. You have not yet assigned any colors or composites. But you are looking at real satellite data of a real place, and that matters.

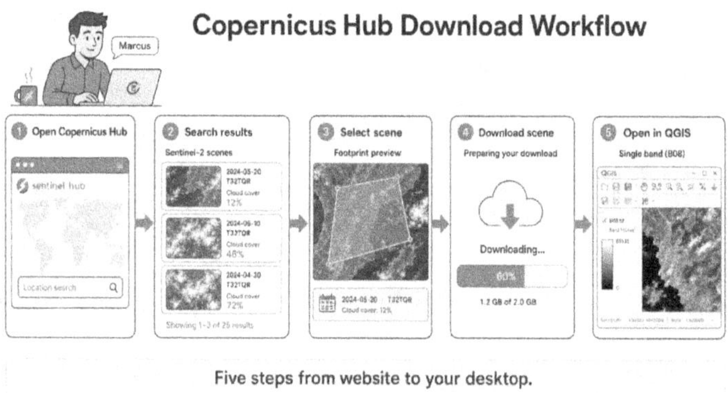

Diagram 9.5: Copernicus Hub Download Workflow

What You Are Actually Looking At

A Sentinel-2 Level-2A product arrives as a folder containing multiple TIFF (Tagged Image File Format) files, each representing one spectral band. You will see files labeled B01 through B12, plus B8A, each corresponding to a different wavelength range. B02, B03, and B04 are the blue, green, and red visible bands. B08 is the broad near-infrared band. B11 and B12 are shortwave infrared. The remaining bands cover coastal aerosol, vegetation red edge, and water vapor.

Each file is a grid of numbers. Each number represents the reflectance value for one 10-meter pixel (or 20-meter or 60-meter, depending on the band) at that location on the ground. The numbers have already been scaled so that 0 means no reflection and 10,000 would represent perfect reflection. Most natural surfaces fall between 500 and 3,000. Open one band file in QGIS and apply a grayscale stretch: you will see land, water, clouds, and urban areas as distinct gray tones, because they reflect the light in that band at different intensities. This is the raw material. Everything else you will do is processing and interpretation applied on top of this foundation.

Week Two: Your First NDVI Map

Week two builds directly on week one. You already have a Sentinel-2 scene on your computer. Now you will calculate NDVI (Normalized Difference Vegetation Index) and look at your familiar place through the lens of vegetation health for the first time.

NDVI uses two bands: the red band (B04) and the near-infrared band (B08). Healthy vegetation absorbs red light for photosynthesis and strongly reflects near-infrared. Bare soil and pavement reflect both bands roughly equally. Water absorbs near-infrared almost completely. The formula compares how much near-infrared a surface reflects relative to how much red it reflects: a high positive number indicates vigorous vegetation, a value near zero indicates bare ground or sparse cover, and a negative value usually indicates water or snow. You do not need to compute the formula manually. QGIS and EO Browser will handle the arithmetic once you tell them which bands to use.

Here are the steps for week two using QGIS:

1. Open QGIS and load your Sentinel-2 scene's B04 (red) and B08 (near-infrared) band files. Drag both TIFF files onto the QGIS canvas.

2. Open the Raster Calculator from the Raster menu at the top of the screen.

3. In the calculator, type the NDVI formula using the layer names QGIS has assigned to your two band files. The formula structure is: (NIR band minus Red band) divided by (NIR band plus Red band). QGIS's Raster Calculator uses a specific syntax for referencing layers, which the QGIS documentation explains clearly with examples.

4. Choose an output file name and location. Click OK. QGIS will calculate the result and add a new layer to your canvas.

5. The new layer will appear in a default grayscale style. Right-click on it in the Layers panel and select Properties, then Symbology.

6. Change the render type from "Singleband gray" to "Singleband pseudocolor." This tells QGIS to apply a color ramp based on the pixel values.

7. Select a diverging color ramp that runs from red or brown (low values) through white (zero) to green (high values). Many users prefer the "RdYlGn" (Red-Yellow-Green) ramp for vegetation maps.

8. Set the minimum value to -0.2 and the maximum to 0.8. This range captures most natural surface types without letting extreme outliers distort the color stretch.

9. Click OK. Your NDVI map now shows your familiar place painted in vegetation-health colors.

What you see may surprise you. Forests appear bright green. Lawns and parks show a lighter green. Agricultural fields vary by crop type and growth stage. Pavement and rooftops appear red or brown. Rivers and ponds appear very dark or negative. Road networks appear as fine red lines cutting through the green. The result is a functional vegetation-health map, not a demonstration or a textbook illustration, but an actual analysis product derived from real satellite data, computed by you, covering a place you know.

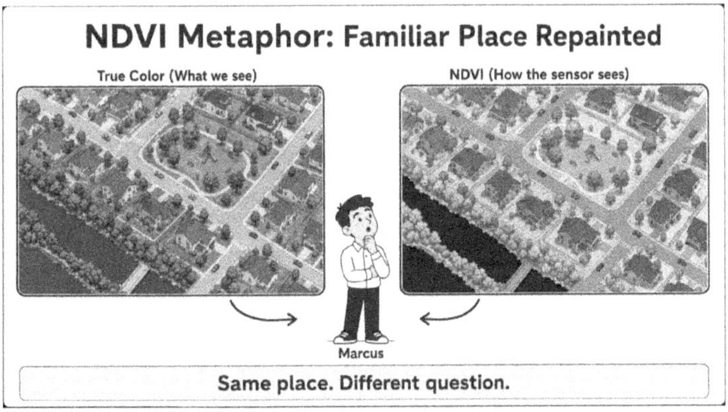

Diagram 9.6: NDVI Metaphor: Familiar Place Repainted

Reading Your NDVI Result

Take a few minutes to read your NDVI map the way a field ecologist would. Find the city park or the forest patch you expected to see. Is the vegetation health uniform, or are there patches of lower greenness within the forest? Lower values within an otherwise healthy woodland can indicate a recent disturbance, a drought-stress hotspot, or simply a clearing you forgot was there. Find the agricultural fields, if any, that are visible. Do different fields show different NDVI values? That variation reflects differences in crop type, irrigation, or growth stage.

Check whether the urban core appears in your scene. The pattern of low NDVI in dense urban areas is one of the most visually striking results of the exercise. Cities read almost entirely as red and brown, with occasional bright-green dots where parks and trees have been preserved. If you live in a city, this pattern makes a powerful personal statement about the distribution of green space. If you are looking at an agricultural region, the pattern makes a different statement: the dominant land use is legible immediately, and the health variation across the landscape is visible in a way that ground-level observation could never provide.

EO Browser offers an alternative approach to this same exercise. If you prefer a browser-based workflow, navigate to your location in EO Browser, select a Sentinel-2 scene from around the same date as the one you downloaded, and click the NDVI preset in the visualization panel. The result will appear instantly, color-coded using a similar red-to-green scale. You can compare the browser result with your QGIS result to verify that both tell the same story. Small differences in color scaling are normal; the underlying pattern should be essentially identical.

Week Three: Your First Change Detection

Week three introduces time. You have a snapshot of your familiar place. Now you will find a second snapshot from a different date and compare the two. This is change detection in its simplest possible form, and the concept from earlier chapters will become visceral and personal when you see it applied to a place you know.

For week three, you will need a second Sentinel-2 scene of the same location, from a time period at least six months different from your first scene. The contrast between seasons produces the most dramatic results for beginners. A summer scene showing peak greenness and a winter scene showing dormant vegetation is an obvious and satisfying comparison. Alternatively, if you know your chosen location has experienced a specific change (a construction project, a wildfire, a flood, a major storm), finding scenes from before and after that event will produce an even more compelling result.

Here are the steps for week three:

1. Return to the Copernicus Data Space Ecosystem and download a second Sentinel-2 L2A scene of the same location, from a different season or a different year. Apply the same cloud-cover filter you used in week one.

2. In QGIS, calculate an NDVI layer for the new scene following the same process as week two. You should now have two NDVI layers: one from each date.

3. Open the Raster Calculator again. This time, subtract the first NDVI layer from the second: new date minus old date. The result is a change

layer in which positive values indicate vegetation has increased, negative values indicate vegetation has decreased, and zero indicates no change.

4. Apply a diverging color ramp to the change layer. Use green for positive change (vegetation gain), red for negative change (vegetation loss), and white or light gray for near-zero values.

5. Set the minimum and maximum to something symmetric, such as -0.3 and +0.3. This prevents extreme values from dominating the color scale.

6. Load the result and examine it against your knowledge of the place.

What you see depends on what changed. Comparing two seasonal scenes will produce a strong seasonal signal: areas that are green in summer and dormant in winter show up as net negative change if your old scene is summer and your new one is winter. Reverse the order, and you see net positive change in the same areas. Both results are correct. Change detection is directional: the sign of the result depends on which date you call "before" and which you call "after."

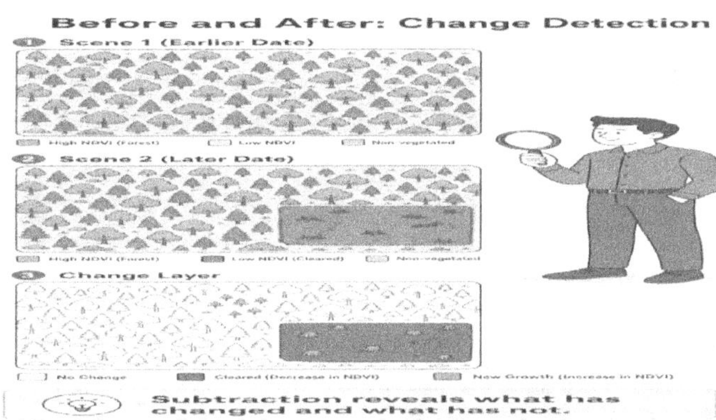

Diagram 9.7: Before and After: Change Detection

Learning from Your Mistakes

Marcus made three predictable mistakes in his first change detection, and you will probably make the same ones. Knowing them in advance saves time.

The first mistake is comparing scenes from different seasons and interpreting the seasonal signal as a real change. A forest that appears to lose half its NDVI value between August and January has not been deforested. The trees have gone dormant. Always compare like-season pairs when looking for non-seasonal change.

The second mistake is ignoring clouds and cloud shadows. Even a 10-percent cloud-cover scene has patches where the sensor saw cloud rather than ground. Those patches will produce spurious change values because the cloud reflectance differs markedly from the ground reflectance beneath it. In your change layer, cloud shadow areas often appear as false negatives and cloud pixels as false positives. You can mask clouds using the quality band (SCL, Scene Classification Layer) included in every Sentinel-2 L2A product. QGIS can apply this mask before you calculate your change layer, and the QGIS documentation has a clear tutorial on the process.

The third mistake is misreading the coordinate reference system. Sentinel-2 scenes are distributed in UTM (Universal Transverse Mercator) projection, and if you stack two scenes from downloads that used different UTM zones, the images will appear offset from each other. QGIS will warn you if you try to combine layers with different projections, and the fix is straightforward: reproject one layer to match the other using the Raster Projections tool. After three or four scenes, handling projections becomes second nature.

Weeks Four Through Eight: A Personal Project

Weeks four through eight are yours. You have the foundational skills. Now you choose a question and answer it. The goal is to produce one real geospatial product: a map, a time-series chart, a before-and-after comparison, or a combination of these, accompanied by a short paragraph of plain-language explanation that a non-specialist could read and understand. Think of it as a one-page geospatial story about a place you care about.

Choosing the right project makes all the difference. The best personal projects have three qualities. First, they concern a place where you have genuine curiosity, because curiosity sustains effort when tools are frustrating. Second, they ask a question that satellite data can actually answer, because not every question has a visible signature from above. Third, they are scoped tightly enough to complete in a few weeks of part-time effort, because scope creep kills momentum.

Good project candidates include:

- Mapping the seasonal greenness cycle of a local park, nature reserve, or urban forest patch and describing what you find in a short paragraph.
- Comparing vegetation health on farmland before and after a drought event that you know occurred in a specific year.
- Documenting the growth of a suburban area by mapping new impervious surface (pavement and rooftops) over a ten-year Landsat time series.
- Tracking the shoreline position of a local beach or river bend over multiple years to quantify erosion or deposition.
- Monitoring a coastal wetland for changes in inundation extent over seasonal or annual time scales.
- Comparing NDVI values between different neighborhoods in your city to visualize the distribution of tree canopy and green space.

Each of these projects uses the skills you built in weeks one through three, applied to a question that matters to you personally. None of them requires code. All of them can be completed with QGIS, Copernicus data, and patience.

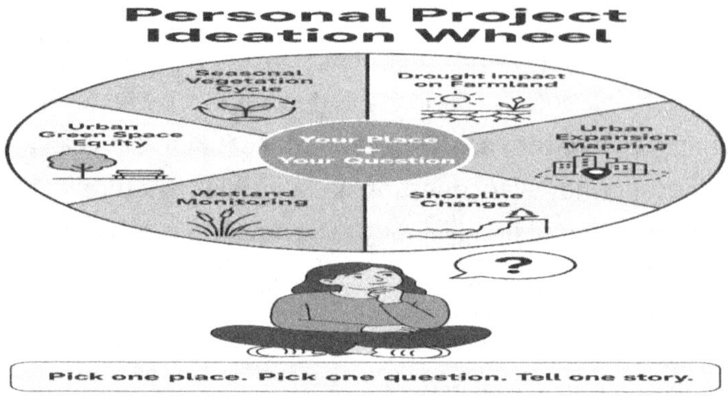

Diagram 9.8: Personal Project Ideation Wheel

Building Your Geospatial Story

A geospatial story is not just a map. A map without context is a picture. A geospatial story adds context: what the place is, what question you asked, what the data shows, and why it matters. Even a paragraph of plain-language explanation transforms a colored raster image into a finding.

Start with a sentence that tells the reader what place you are looking at and why you chose it. Follow with a sentence describing the time period your data covers. Then describe the main pattern you see in plain English, the way you would describe it to a friend who cannot see the map. Use specific language: "the western side of the preserve shows consistently higher NDVI values than the eastern side, which may reflect the soil moisture gradient between the two slopes." Finally, note one thing you do not know from the data alone, one question that your map raises but cannot answer by itself. That honest acknowledgment of limitation is what separates a geospatial story from a sales pitch.

When your one-page story is finished, share it. Post it in the OSGeo community forum. Share it on LinkedIn using the hashtag geospatial practitioners use to share work. Email it to the local park service or city planning department that oversees the place you studied. You may be surprised by the response. Real organizations with real decisions to make are often delighted when a motivated newcomer arrives with actual data and a clear question. The skills you have built in eight weeks are not hobbyist skills. They are the same skills at the foundation of professional practice.

Common Tools for Visualization and Storytelling

QGIS has a full print layout composer for designing print-quality maps with a legend, scale bar, north arrow, and title. It exports to PDF and image formats. Take time with the map design: a clear title, a clean color scheme consistent with your NDVI ramp, labeled features, and a short caption make the difference between a map that communicates and one that confuses.

For time-series work, EO Browser has a built-in animation export that creates a simple GIF (Graphics Interchange Format) showing your location changing through time. These animations communicate seasonal or multi-year change immediately to non-specialist audiences. Export the animation, add a caption, and you have a compelling one-minute presentation.

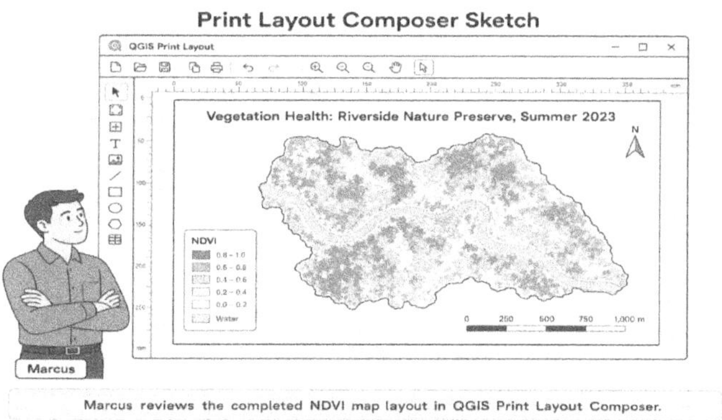

Diagram 9.9: Print Layout Composer Sketch

Where to Go Next: Certifications, Communities, Careers

Eight weeks of personal practice puts you at the beginning of a field, not the end of a textbook. The geospatial profession has a genuine skills shortage at the analyst level across multiple sectors: government, agriculture, climate, urban planning, humanitarian response, and defense. If you find that the work you did over these eight weeks energized you rather than exhausted you, there are clear paths forward, and most of them are more accessible than you might expect.

Professional Certifications

The GISP (GIS Professional) credential, administered by the GIS Certification Institute, is the most widely recognized certification for GIS practitioners in North America. It requires documented experience and a passing knowledge assessment. The GISP is not a beginner's credential, and eight weeks of practice is not enough to qualify. But the competency framework it describes (data management, analysis, cartography, project management, programming) is an excellent guide to organizing your long-term skills development.

OGC (Open Geospatial Consortium) offers training and certification programs focused on geospatial standards, interoperability, and web services. OGC credentials are particularly relevant for government data infrastructure, environmental monitoring, or enterprise GIS deployments where data must communicate across organizational boundaries. The training catalog includes self-paced online courses and instructor-led workshops, many at low or no cost.

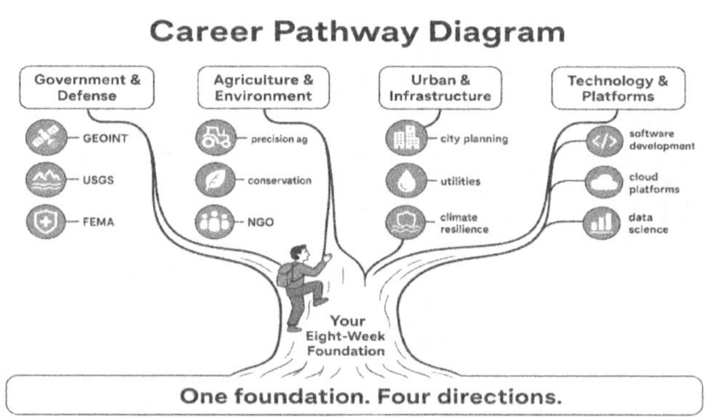

Diagram 9.10: Career Pathway Diagram

The GEOINT Community

GEOINT (Geospatial Intelligence) is the professional discipline that integrates satellite imagery, geospatial data, and intelligence analysis. The GEOINT Symposium, organized by USGIF (the United States Geospatial Intelligence Foundation), is the largest professional gathering in the field, held annually and attracting thousands of practitioners, vendors, researchers, and government officials. USGIF also publishes the GEOINT Body of Knowledge and maintains a professional network that welcomes people at all career stages.

Even if national security work is not your goal, the GEOINT community is worth getting to know. It produces some of the most sophisticated public thinking about the future of Earth observation, data ethics, privacy, and satellite governance. The conversations happening in that community now will shape the tools and norms of the entire geospatial field for the next decade.

Earth on AWS and Open Data

AWS (Amazon Web Services) hosts the Registry of Open Data on AWS, a catalog of publicly available datasets that can be analyzed directly in the cloud. The Earth observation section includes the full Landsat and Sentinel-2 archives, MODIS data, global elevation models, and hundreds of derived products. For anyone who wants to work at scale beyond a personal laptop, cloud access changes the economics of geospatial analysis entirely: you pay only for compute time, and data transfer is free.

AWS also maintains beginner tutorials for Earth observation data in the cloud. Pairing those tutorials with your QGIS and EO Browser skills creates a toolkit that spans from desktop exploration to cloud-scale analysis, covering the range of what most entry-level and mid-level analysts actually do.

Communities That Welcome Beginners

Three communities are particularly welcoming to newcomers and consistently produce valuable informal learning.

The OSGeo (Open Source Geospatial Foundation) community is organized around the free and open-source geospatial tools, including QGIS, GDAL (Geospatial Data Abstraction Library), GeoServer, and others. OSGeo holds an annual conference called FOSS4G (Free and Open Source Software for Geospatial), where hundreds of talks are recorded and made freely available online. The OSGeo community forums and mailing lists are friendly, technically deep, and genuinely responsive to beginner questions.

The r/gis subreddit is a large, active community where practitioners at all levels post questions, share work, and discuss news. The culture is helpful rather than gatekeeping, and a well-described beginner question typically receives several substantive answers within a few hours. For quick troubleshooting ("Why is my QGIS raster calculator showing all zeros?"), it is often the fastest resource available.

The geospatial practitioner community on LinkedIn and Twitter/X, organized around hashtags like #geospatial, #remotesensing, and #EarthObservation, is active and welcoming. Posting your first NDVI map or personal project result, along with a short explanation, is an effective way to introduce yourself and receive feedback from practitioners who remember what it felt like to see their first result.

Diagram 9.11: Community Ecosystem Map

A Note on Careers

The geospatial skills shortage is real. Employers across agriculture, emergency management, climate research, urban planning, and national security report difficulty finding candidates who combine spatial thinking with domain knowledge. An agronomist who can communicate a satellite-based crop assessment is more valuable than either an agronomist or a remote sensing analyst working alone. An urban planner who can produce a Landsat heat-island map is more effective than one who must wait for a specialist. A journalist who can independently verify a satellite-imagery claim is providing a service that is increasingly rare and valued.

You do not need to change careers to use these skills. Bring them to the career you already have.

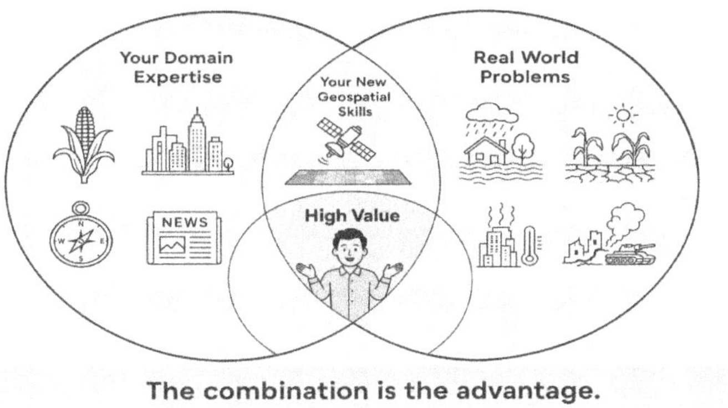

Diagram 9.12: Skills Integration Diagram

Takeaway

Eight weeks is not a long time. It is two months. It is roughly 40 to 60 hours of focused practice, spread across evenings and weekends, or concentrated into a series of workdays. In that time, the path Marcus followed, and the path this chapter has laid out, takes you from zero to holding a real result: a map you made yourself, from data you downloaded yourself, answering a question you chose yourself. That result may not yet be sophisticated enough to present in a professional context. But it is real, and it is yours, and it is built on the same conceptual foundation that every experienced geospatial analyst carries.

The concepts in the earlier chapters of this book are no longer abstractions after you complete these exercises. Spectral bands are the individual files sitting in your Sentinel-2 folder. NDVI is the formula you typed into the Raster Calculator and the green-to-red image that appeared on your screen. Change detection is the subtraction you performed and the red patches that appeared over the construction site. Time series is the EO Browser animation you exported and watched in a loop. The vocabulary that once felt foreign is now anchored to experience.

The field of geospatial intelligence is not a closed club. It is a discipline in expansion, adding new satellites, sensors, open-data policies, and communities every year. The tools described in this chapter are the same tools used by analysts at humanitarian organizations responding to floods, researchers tracking deforestation, planners designing climate-resilient cities, and scientists monitoring glaciers from the other side of the planet. You are equipped to join those conversations, to ask informed questions, to evaluate claims critically, and, if you choose, to contribute analysis of your own. The planet is observable. You now know how to observe it.

10 Closing

The View from Here

Somewhere over the Pacific right now, a small satellite the size of a shoebox is sweeping a strip of ocean. Its sensor is reading reflected sunlight in wavelengths the human eye cannot see. The data it collects will reach a ground station in a few minutes, move through a processing pipeline in a few more, and arrive in a cloud storage bucket where an analyst or an algorithm is waiting. By tomorrow morning, that strip of data may be part of a fishing-fleet alert, an ocean-temperature record, a methane baseline, or a ship-tracking product delivered to a customer who will never know a satellite was involved. The whole sequence takes less time than it takes most of us to commute to work.

You now understand almost every step in that sequence. That is not a small thing.

Nine chapters ago, a satellite image was probably something you encountered on the news and accepted on faith. You saw the orange smoke plume over a California ridge, or the before-and-after of a flood-scoured city, or the infrared map of a contested border, and you had no particular reason to doubt it and no particular framework for interrogating it. That passivity was understandable. Nobody had explained the machinery.

Now you have the machinery. You know that the orange in that smoke image was a deliberate color choice, not a photograph. You know that the before-and-after flood map was built from two scenes with different capture times, sun angles, and possibly different sensors, and that a responsible analyst validated the change with something other than the image itself. You know that the infrared border map was almost certainly a false-color composite, that synthetic aperture radar (SAR, the sensor that uses its own radio pulse like a bat's sonar) can see through clouds and at night, and that the commercial providers who sell those images sell them to governments, journalists, and hedge funds alike. You know what questions to ask.

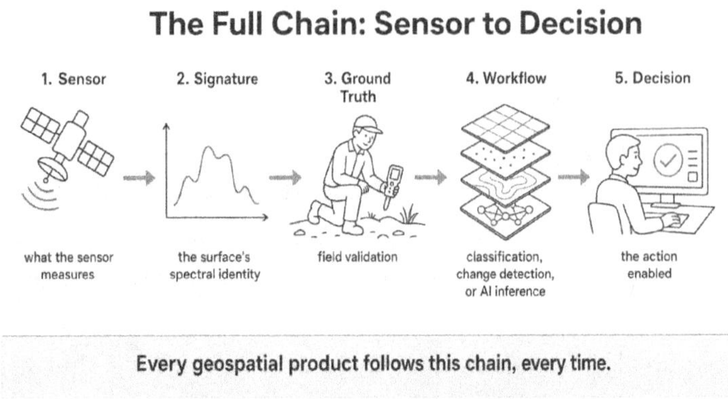

Diagram C.1: The Full Chain: Sensor to Decision

What the Nine Chapters Actually Said

The nine chapters of this book were never really about nine separate topics. They were nine views of the same thing, each approached from a different angle.

Chapter 2 introduced the electromagnetic spectrum as a piano keyboard: visible light is a handful of keys near the middle, and the sensors that matter most to analysts reach far to the left and right of what the eye can see. That keyboard is not a detail. It is the foundation of everything. Every sensor decision in every subsequent chapter was a choice about which keys to press and why.

Chapter 3 showed you that a satellite is a procurement decision as much as it is a piece of technology. Spatial resolution, revisit time, spectral range, and cost are four levers, and every imagery program is a different trade-off among them. Knowing the trade-offs makes you a sharper consumer, whether you are reviewing a vendor's pitch or reading a news story citing satellite evidence.

Chapters 4 and 5 established the two most important mental habits in the field: skepticism about single images and respect for time. A single image is a snapshot. Every surface it shows was illuminated by a sun at a specific angle, filtered by a specific atmosphere, captured by a sensor with a specific calibration on a specific morning. Change detection and time-series analysis are where the real power lives. The planet has rhythms, and every modern commercial Earth-observation product is built on an engine that continuously tracks them. The EKG metaphor from Chapter 5 holds: one heartbeat tells you almost nothing; a thousand heartbeats across six months tell you whether the patient is well.

Chapter 6 took AI off its pedestal. Geospatial AI is not magic, and it is not infallible. It is a pattern-recognizer trained on labeled examples, subject to every bias in those examples and every mismatch between the training and deployment environments. The best models in the world still need a human in the loop during validation. That will not change. What will change, and is already changing fast, is how much work the model can do before a human needs to look. Foundation models for Earth observation (Prithvi, Clay, SatMAE, and their successors) are compressing the gap between a raw satellite archive and a usable analytic product. The analyst's job is shifting from computation to judgment.

Chapter 7 mapped the applications—agriculture, urban planning, climate science, insurance, disaster response, energy, shipping. The same conceptual pattern runs through all of them: choose a sensor, find the spectral signature or temporal pattern that encodes what you care about, validate with ground truth, build a workflow, and deliver a decision. The pattern is portable. Once you see it, you will recognize it everywhere.

Chapter 8 named the dual-use reality without flinching. The constellation that monitors deforestation can monitor troop movements. The algorithm that counts fishing vessels can count naval vessels. The SAR image documenting a humanitarian disaster can also document a military facility. Geospatial intelligence has always lived at this intersection, from the CORONA film capsules dropped from orbit in the early 1960s to the commercial imagery analysts who published open-source intelligence about the invasion of Ukraine in real time in

2022. Informed citizenship now requires at least a working acquaintance with this technology.

Chapter 9 opened the door to practice. The tools are free. The data is free. The training is free. Marcus's 60-day starter path was not a hypothetical. It is available to anyone with a laptop and a browser.

You Are Geospatially Literate

There is a specific, transferable skill set you now carry. It is worth naming precisely.

You can read a satellite-derived news story and ask the right questions: What sensor captured this? What processing was applied? What is the capture date relative to the event being described? Is this a true-color image or a false-color composite, and does the story say so? Was any ground truth provided, or is this analysis resting entirely on the image?

You can evaluate a geospatial vendor without being dazzled by a demo. When a vendor says their model achieves 90% accuracy, you know to ask: 90% on which validation dataset, in which biome, at which season, with what resolution imagery? You know that accuracy is always conditional.

You can walk into Google Earth Engine, USGS Earth Explorer, or Copernicus Open Access Hub without feeling lost. You know what a GeoTIFF is, what the metadata fields mean, and what the difference between a Level 1 and Level 2 product is. You know the first things to check are the capture date and cloud cover.

You can participate in conversations about climate, agriculture, defense, and urban planning as someone who understands the instruments behind the conclusions. That is not a trivial capability. Most people in those conversations do not have it.

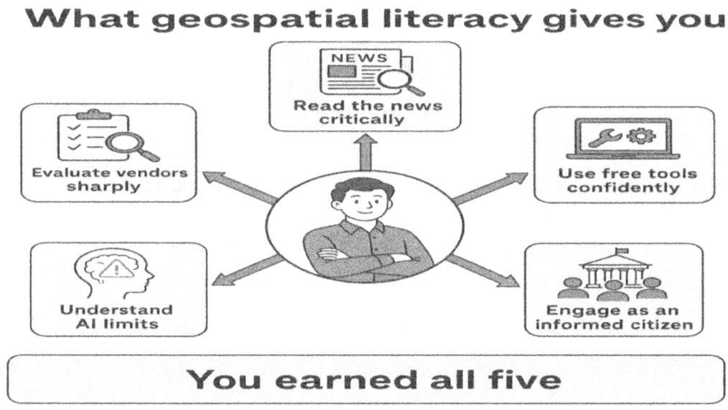

Diagram C.2: The Geospatially Literate Reader

The Next Five to Ten Years

The field is accelerating. It is worth knowing what is coming, both because it is genuinely exciting and because some of it deserves the same critical attention you now know how to apply.

The most important near-term development is the maturation of foundation models for Earth observation. Models like Prithvi and Clay, mentioned in Chapter 6, are pretrained on hundreds of millions of satellite scenes. They can be adapted to new tasks, new geographies, and new sensor types with far less labeled training data than classical machine learning required. In practical terms, this means that applications that previously required months of data labeling and model training will eventually require days. The barrier to entry for a new geospatial product will drop substantially. That is good for the many organizations, especially in the developing world, that currently cannot afford custom model development. It also raises new questions about who is watching what, at what scale, with what oversight.

Commercial constellations will continue to multiply. Planet already images most of the Earth's land surface every day. In the next decade, higher-resolution daily imaging will become available at lower cost, and the temporal gap between an event and the availability of an image will narrow from hours to minutes for some providers. Hyperspectral imagery, which captures dozens or even hundreds of narrow spectral bands rather than a handful of broad ones, will move from specialty missions to routine commercial products. The piano keyboard from Chapter 2 will have far more keys available for analysis.

Open data will expand. The European Space Agency's Copernicus program is already one of the most consequential scientific data-sharing decisions in history. NASA's commitment to open access for Landsat, MODIS, and an expanding list of missions has set a standard. Other space agencies and commercial providers are moving, if unevenly, toward more openness. The direction is clear even if the pace is not.

Open-source intelligence will deepen. The combination of commercial imagery, social media geolocation, and citizen analyst communities has permanently changed who can know what about events on the ground. That has already reshaped conflict journalism, human rights documentation, and disaster response coordination. The trend will continue. The skills you now have are exactly the skills a thoughtful citizen analyst needs.

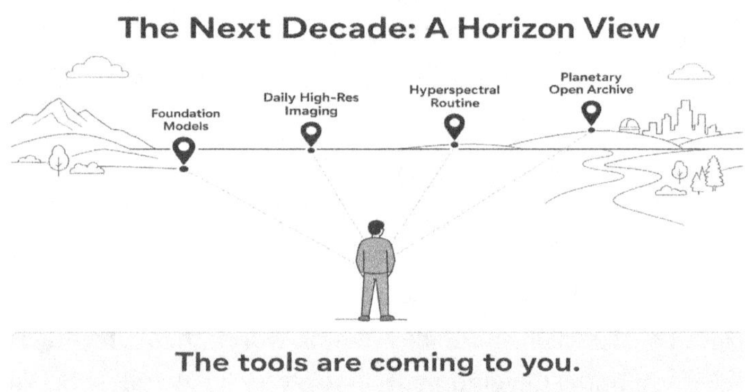

Diagram C.3: The Next Decade: A Horizon View

What Informed Citizens Should Watch

Being geospatially literate does not mean being uncritical of the field. A few things deserve your attention over the coming years.

Persistent surveillance is the most significant civil-liberties question the commercial constellation era has raised. When any location on Earth can be imaged multiple times a day, the privacy implications extend well beyond national security. Protests, places of worship, border crossings, and private residences are all visible from orbit. The legal frameworks governing what commercial providers can sell, to whom, and at what resolution are still evolving. Resolution limits enforced by the Kyl-Bingaman amendment in the United States are one example of a policy instrument applied to imagery. The conversation about whether such limits are adequate for a world of daily half-meter imaging is just beginning.

Training data bias in geospatial AI has real-world consequences. A model trained primarily on North American and European urban environments and then deployed in West African or Southeast Asian contexts may produce confident wrong answers. Those wrong answers may feed into insurance risk models, humanitarian aid allocations, or infrastructure assessments in the places that can least afford the errors. Asking about training data geography is not a technical nitpick. It is a basic due diligence question that any informed customer should ask.

The open-source intelligence (OSINT) ecosystem is powerful and sometimes wrong. The same democratization that has enabled journalists and human rights organizations to document atrocities has also enabled misidentification, deliberate disinformation, and the rushed publication of analysis that a slower process would have caught. Reading OSINT with the same critical habits you now apply to any satellite-derived claim, checking the sensor, the date, the processing, and the validation, is the right posture.

Diagram C.4: Three Things to Watch

One Last Moment at Pleiades

It is a Tuesday afternoon at Pleiades Geospatial Services, and Marcus is showing a new hire a screen.

The screen shows a time-series animation: forty years of Landsat imagery of a lake in East Africa. The shoreline moves. In the early frames, the lake is large. In the middle frames, it contracts. By the final frames, the lake is smaller than it has been at any point in the record, and the land that was once underwater is pale and cracked. It takes about twenty seconds to watch. It says something that no single image could say.

The new hire asks what caused it. Marcus says there are several hypotheses, all supported by different lines of evidence, and that the team is building a workflow to distinguish them. He does not say it is simple. He does not say a model already knows the answer. He says the data is there, the tools are free, and the question is good. Then he opens a new browser tab and starts.

Amara, walking past, glances at the screen and nods. She has seen a hundred lakes like this one. Each time, she says, the image does not tell you everything. But it does tell you something, and something is what you build on.

That is the whole field, stated plainly. The sensor captures what it can. The signature indicates which surface you are looking at. The time series tells you how it is changing. The workflow translates the change into something a decision-maker can use. The decision-maker is a human being who carries the weight of getting it right.

You now understand each step in that chain. You know its strengths and its limits. You know the questions to ask and where to find the tools to answer them.

The satellite is still up there, sweeping its strip of ocean. The data is still arriving. What you do with it from here is up to you.

11 Glossary

Every term you met in this book, defined in plain English the way a friendly colleague would explain it over coffee. Alphabetical by term.

A

AIS (Automatic Identification System). A radio signal that ships broadcast continuously to identify themselves, share their position, and report their heading and speed. Combined with SAR satellite imagery, AIS lets analysts spot vessels that have quietly switched off their transponder to avoid being tracked.

ARD (Analysis Ready Data). Satellite imagery that has already been atmospherically corrected, georeferenced, and tiled so an analyst can start working with it immediately. The prep work is done before it arrives; you open the file and start analyzing rather than cleaning.

ARSET (Applied Remote Sensing Training). A free NASA training program that teaches scientists, resource managers, and decision-makers how to apply satellite data to real problems. ARSET offers online and in-person courses on topics from flood mapping to air quality monitoring.

ASPRS (American Society for Photogrammetry and Remote Sensing). A professional association that sets technical standards for remote sensing, photogrammetry, and geospatial imaging. ASPRS also administers the Certified Photogrammetrist credential and publishes the journal Photogrammetric Engineering and Remote Sensing.

Atmospheric correction. The process of mathematically removing the haze, scattering, and absorption effects of the atmosphere from a raw satellite image, so the reflectance values represent the ground surface rather than a mixture of surface and air. It is the difference between what the sensor measured and what was actually there.

B

BlackSky. A commercial Earth observation company that operates a constellation of small satellites designed for frequent revisit and rapid tasking. BlackSky emphasizes analytics and event monitoring at key sites rather than large-area mapping.

BRDF (Bidirectional Reflectance Distribution Function). A mathematical description of how a surface reflects light in different directions depending on the angle of the sun and the viewing angle of the sensor. BRDF corrections help ensure that the same patch of ground looks the same brightness no matter when or from what angle it was imaged.

C

Capella Space. A commercial SAR satellite company based in the United States. Capella's radar sensors can collect high-resolution imagery at night and through clouds, making them useful for persistent monitoring of ports, facilities, and disaster zones regardless of weather.

Change detection. An analytical process that compares satellite images of the same area taken at different times to identify changes. Urban growth, deforestation, flood extent, and hurricane damage are all classic change-detection applications.

Classification (supervised). A machine-learning technique in which an analyst provides labeled example pixels (this pixel is forest, this one is water) and then trains an algorithm to assign every other pixel in the image to one of those categories. The algorithm learns from human-provided examples.

Classification (unsupervised). A technique in which an algorithm groups pixels by their spectral similarity without any human labels, producing clusters that the analyst then interprets and names. Useful when you do not know in advance what land-cover types are present.

Clay. An open-source geospatial foundation model trained on a large corpus of multi-spectral satellite imagery. Clay produces general-purpose embeddings that researchers and engineers can fine-tune for specific tasks without building a model from scratch.

CNN (Convolutional Neural Network). A type of deep learning model that is particularly good at finding spatial patterns in images. CNNs power many object-detection and land-cover classification systems in geospatial intelligence, learning to recognize features like roads, buildings, and ships from labeled training data.

COG (Cloud-Optimized GeoTIFF). A GeoTIFF file arranged internally so that clients can request and download only the portion of the image they need from a cloud server. A COG lets a web browser show a small preview without downloading the full multi-gigabyte file.

D

Dual-use. The property of a technology or piece of information that can serve both civilian and military purposes. Satellite imagery is the classic dual-use technology: the same image of a port can help a shipping company plan logistics and help an adversary plan an operation.

E

ESA (European Space Agency). The intergovernmental space agency of Europe, responsible for the Copernicus program (including Sentinel-1 and Sentinel-2), the Earth Observation data archive at the Copernicus Open Access Hub, and dozens of other Earth observation and science missions.

EVI (Enhanced Vegetation Index). A vegetation index that builds on NDVI by correcting for atmospheric interference and for soil background signals. EVI is particularly useful in dense tropical forests where NDVI tends to saturate and lose its ability to distinguish different levels of canopy health.

F

Foundation model. A large AI model trained on a broad, diverse dataset that can then be adapted for many downstream tasks with relatively little additional training. In geospatial contexts, foundation models trained on satellite imagery are beginning to replace task-specific models for common analysis jobs.

FOSS4G (Free and Open Source Software for Geospatial). Both a description of a category of software (open-source GIS tools like QGIS, GDAL, and PostGIS) and the name of the annual international conference that brings the open geospatial community together. FOSS4G events take place globally and regionally.

G

GeoJSON. A lightweight text-based format for representing geographic features (points, lines, polygons) and their attributes. GeoJSON is widely used in web mapping applications because it is easy for browsers and APIs to read and write.

GeoTIFF. A standard raster image format that stores geographic metadata (coordinate reference system, geographic extent, pixel resolution) inside the image file itself. GeoTIFF is the most common file format for distributing satellite imagery.

GHGSat. A commercial company that operates a constellation of satellites designed specifically to detect and measure greenhouse gas emissions, particularly methane and carbon dioxide, from individual industrial facilities. GHGSat data helps companies and regulators identify and quantify emissions at the source level.

GISP (Geographic Information Systems Professional). A professional certification administered by the GIS Certification Institute recognizing demonstrated competency in GIS practice, ethics, and education. GISP is widely recognized by employers in government and private-sector GIS roles.

GOES (Geostationary Operational Environmental Satellite). A series of weather satellites operated by NOAA that orbit at geostationary altitude (approximately 35,786 kilometers), hovering over the same spot on the equator continuously. GOES captures wide-area imagery of weather systems every few minutes and is the backbone of American operational weather forecasting.

Ground truth. Data collected on the ground, through field surveys, GPS measurements, or other direct observation, used to verify and validate what a satellite or model tells you. If a model predicts that a field is corn, ground truth confirms whether it actually is.

H

Hague Code of Conduct. An international political commitment, not a legally binding treaty, in which subscribing states agree to transparency about their ballistic missile and space-launch programs. It is often cited in discussions about norms and confidence-building in the use of space.

HDF5 (Hierarchical Data Format 5). A file format designed to store large, complex scientific datasets efficiently. Many satellite data products from NASA and NOAA are distributed in HDF5 because it can hold multiple data layers, metadata, and different spatial resolutions in a single file.

Himawari. A series of geostationary weather and Earth observation satellites operated by the Japan Meteorological Agency, positioned over the Asia-Pacific region. Like GOES for the Americas, Himawari provides near-continuous high-frequency imagery of clouds, storms, and other atmospheric phenomena.

HOTOSM (Humanitarian OpenStreetMap Team). An organization that coordinates volunteers to add detailed geographic data to OpenStreetMap in response to disasters and humanitarian crises. When a cyclone or earthquake strikes, HOTOSM volunteers can produce a detailed map of affected areas within hours using satellite imagery.

I

ICEYE. A Finnish company operating one of the largest commercial SAR satellite constellations. ICEYE satellites are compact and numerous, enabling frequent revisit of the same location with radar imaging regardless of cloud cover or darkness.

IoU (Intersection over Union). A metric used to evaluate how accurately a model detects or segments objects in an image. IoU compares the area of overlap between a model's prediction and the true boundary to the total area covered by both, with a score of 1.0 meaning a perfect match.

J

JAXA (Japan Aerospace Exploration Agency). Japan's national space agency, responsible for Earth observation satellites such as the ALOS series (which carries L-band SAR sensors) and climate and oceanography missions. JAXA data products are widely used in forestry, disaster response, and ocean research.

K

KML (Keyhole Markup Language). An XML-based file format for representing geographic data, originally developed for Google Earth. KML files describe points, lines, polygons, and imagery overlays in a way that applications like Google Earth can display and navigate interactively.

Kyl-Bingaman Amendment. A provision of United States law that historically required commercial satellite operators to limit the resolution of imagery of Israel and the Palestinian territories sold to the public to match the best commercially available resolution from non-U.S. sources. The restriction was significantly relaxed in 2020 as foreign competition made the original limit obsolete.

L

Landsat. A series of Earth observation satellites jointly managed by NASA and the USGS, with imagery dating continuously back to 1972. Landsat is the longest continuous record of the Earth's land surface from space and is foundational to studies of deforestation, urban growth, agricultural change, and glacier retreat.

Level 1. A satellite data product that has been processed minimally, typically to correct for instrument artifacts and assign geographic coordinates to each pixel. Level 1 data is the starting point for further processing; the analyst still needs to do atmospheric correction and other steps before analysis.

Level 2. A satellite data product that has been atmospherically corrected, delivering surface reflectance or surface temperature rather than raw radiance. Analysts generally use Level 2 products for vegetation indices and land-cover mapping.

Level 3. A satellite data product that has been spatially or temporally composited, such as a monthly average or a mosaic of cloud-free observations. Level 3 products trade some raw detail for greater spatial or temporal coverage.

LiDAR (Light Detection and Ranging). A remote sensing technique that fires rapid pulses of laser light at the ground and measures how long each pulse takes to return. The result is a dense, precise three-dimensional point cloud of the surface and any objects on it, used to measure terrain elevation, tree canopy height, and building structure.

M

Major TOM. A large open dataset designed to train geospatial foundation models, providing diverse global satellite imagery samples that cover a representative range of land-cover types and geographic regions. The name reflects the aspiration to provide a comprehensive "major" data corpus.

Maxar Technologies. A leading commercial high-resolution Earth observation company operating satellites including WorldView-2, WorldView-3, and WorldView Legion. Maxar imagery, with resolutions down to 30 centimeters, is widely used in defense, intelligence, mapping, and infrastructure monitoring.

MethaneSAT. A satellite operated by the Environmental Defense Fund designed specifically to detect and measure methane emissions from the oil and gas sector at a regional scale. MethaneSAT aims to provide the data needed to hold industries and governments accountable for emission reductions.

Microwave (spectrum)—the portion of the electromagnetic spectrum used by radar and passive microwave sensors. Microwave energy passes through clouds and is not blocked by darkness, making microwave-based sensors including SAR and passive radiometers essential for all-weather, day-or-night Earth observation.

MODIS (Moderate Resolution Imaging Spectroradiometer). A sensor aboard NASA's Terra and Aqua satellites that has been collecting multi-spectral imagery of Earth's entire surface every 1 to 2 days since 1999. MODIS captures 36 spectral bands at resolutions from 250 meters to 1 kilometer and is central to global fire, vegetation, and ocean monitoring.

MSI (Multispectral Instrument). The primary sensor aboard ESA's Sentinel-2 satellites. MSI captures 13 spectral bands at resolutions of 10 to 60 meters, enabling detailed land-cover mapping, vegetation health assessment, and coastal monitoring.

N

NAIP (National Agriculture Imagery Program). A USDA program that acquires high-resolution (1-meter) aerial imagery of the contiguous United States during the growing season. NAIP imagery is free, publicly available, and widely used by farmers, planners, and conservation agencies.

NASA (National Aeronautics and Space Administration). The United States federal agency responsible for the civilian space program, aeronautics research, and a large portfolio of Earth-observing satellites including Landsat (jointly with USGS), MODIS/Terra/Aqua, GOES (in partnership with NOAA), and the ICESat ice-sheet monitoring mission.

NBR (Normalized Burn Ratio). A spectral index that uses near-infrared and shortwave infrared bands to identify burned areas and assess fire severity. High pre-fire and low post-fire NBR values indicate severe burning; comparing pre- and post-fire NBR images produces a burn severity map.

NDVI (Normalized Difference Vegetation Index). The most widely used vegetation index in remote sensing, calculated from the difference between near-infrared and red reflectance values divided by their sum. Healthy green vegetation absorbs red light and strongly reflects near-infrared light, producing high NDVI values. Bare soil, water, and dead vegetation produce low values.

NDWI (Normalized Difference Water Index). A spectral index that highlights water features using green and near-infrared bands. NDWI is used to map the extent of surface water, monitor reservoir levels, and detect flooding. Positive values generally indicate water; negative values indicate vegetation or land.

NetCDF (Network Common Data Form). A self-describing file format widely used for multidimensional scientific data such as time series of gridded weather, ocean, or climate model output. NetCDF files carry their own metadata, making them easier to share and interpret across different software tools.

NGA (National Geospatial-Intelligence Agency). The United States federal agency responsible for geospatial intelligence (GEOINT). NGA produces maps, imagery analysis, and geospatial data for defense, intelligence, and navigation purposes, and also manages the GPS ground control segment.

NIR (Near-Infrared). The portion of the electromagnetic spectrum just beyond the red end of visible light, invisible to the human eye but captured by most multispectral satellite sensors. NIR is strongly reflected by healthy green vegetation, making it the key band in vegetation health indices like NDVI.

NRO (National Reconnaissance Office). The United States agency that designs, builds, and operates the classified reconnaissance satellites used by the intelligence community and military. NRO satellites collect imagery, signals intelligence, and other data; the specific capabilities of operational systems are classified.

O

Object detection. A computer vision technique that identifies and locates specific objects (ships, aircraft, buildings, vehicles) within an image by drawing a bounding box around each instance. Object detection is widely used in geospatial intelligence for counting assets and monitoring activity at specific facilities.

OGC (Open Geospatial Consortium). An international standards organization that develops open standards for geospatial data and services, including WMS (Web Map Service), WFS (Web Feature Service), and the GeoPackage format. OGC standards ensure that different GIS software systems can share and display data interchangeably.

Open Skies Treaty. A multilateral arms control agreement, originally signed in 1992, that allowed signatory nations to conduct unarmed surveillance flights over each other's territory. The United States withdrew in 2020, and Russia withdrew in 2021, effectively ending the treaty for its two largest participants.

OSGeo (Open Source Geospatial Foundation). A non-profit organization that supports and promotes the development of open-source geospatial software including QGIS, GDAL, GeoServer, and PostGIS. OSGeo hosts the FOSS4G conference series.

OSM (OpenStreetMap). A free, editable map of the world built by a global community of volunteers. OSM is the largest open geographic database and is widely used as a basemap, for training machine learning models, and as a source of building and road data in humanitarian response.

OSINT (Open-Source Intelligence). Intelligence gathered from publicly available sources including news, social media, commercial satellite imagery, shipping databases, and flight tracking data. Much modern geospatial analysis is built on OSINT, which does not require classified access or covert collection.

P

Phenology. The study of recurring natural events and their timing, such as when crops emerge, when forests leaf out, and when snow melts. Satellite time series reveal phenological patterns and shifts at continental scale, enabling early detection of drought stress, invasive species, and climate-driven changes in growing seasons.

Planet Labs. A commercial Earth observation company operating the world's largest constellation of small optical satellites (Doves), which collectively image Earth's entire landmass every day. Planet also operates higher-resolution SkySat satellites for targeted tasking.

Prithvi. A geospatial foundation model developed jointly by NASA and IBM, trained on years of Harmonized Landsat Sentinel-2 data. Prithvi is designed to be fine-tuned for tasks like flood mapping, crop segmentation, and wildfire burn area estimation.

R

Random forest. A machine learning algorithm that builds many decision trees on random subsets of the training data and combines their predictions, reducing overfitting and producing more reliable classifications. Random forest has been a workhorse of land-cover classification in remote sensing for more than two decades because it handles noisy, high-dimensional spectral data well.

S

SAR (Synthetic Aperture Radar). A type of active sensor that transmits microwave pulses toward the Earth and records the energy reflected. Because SAR uses its own energy source, it works at night, and because microwaves penetrate clouds, SAR provides all-weather imaging capability that passive optical sensors cannot match.

SatMAE. A self-supervised foundation model trained on multi-spectral satellite imagery using a masked autoencoder approach, where the model learns to reconstruct hidden portions of an image. SatMAE learns rich spectral and spatial representations that transfer well to downstream classification and segmentation tasks.

Segmentation. A more precise form of image analysis than object detection, in which a model assigns a class label to every individual pixel rather than drawing boxes around objects. Segmentation produces detailed outlines of buildings, fields, and water bodies, making it well suited for area measurement and boundary delineation.

Sentinel-1. An ESA satellite mission carrying a C-band SAR sensor. Sentinel-1 acquires radar imagery of the entire Earth approximately every six to twelve days, supporting applications in flood mapping, sea-ice monitoring, ship detection, and ground deformation tracking.

Sentinel-2. An ESA satellite mission carrying the Multispectral Instrument (MSI). Sentinel-2's pair of satellites revisit most of the Earth's land surface every five days at 10-meter resolution in visible and near-infrared bands, making it the backbone of European and global land-monitoring programs.

SkyFi. A satellite tasking marketplace that lets customers order imagery from multiple commercial satellite operators through a single platform. SkyFi aggregates capacity from partners including optical and SAR satellite operators, providing access to a range of resolutions and revisit frequencies.

Spectral signature. The characteristic pattern of reflectance or emission values across multiple wavelengths that identifies a material. A healthy pine forest has a distinct spectral signature from corn, asphalt, and snow, allowing analysts to distinguish materials by how they interact with light across the spectrum.

SWIR (Shortwave Infrared). A portion of the infrared spectrum, roughly 1,000 to 2,500 nanometers, captured by sensors such as Landsat 8, Sentinel-2, and WorldView-3. SWIR is particularly sensitive to soil moisture, vegetation water content, burn scars, and mineral composition, giving analysts information that visible bands cannot provide.

T

Thermal infrared. Radiation emitted by objects as heat rather than reflected from the sun, detectable by sensors sensitive to wavelengths of roughly 3 to 14 micrometers. Thermal infrared imagery reveals surface temperatures, enabling applications like detecting volcanic activity, monitoring urban heat islands, and tracking wildfire fronts.

Time series. A sequence of images of the same area acquired at regular intervals over weeks, months, or years. Analyzing a time series reveals trends and patterns that a single image cannot, such as seasonal crop cycles, gradual land subsidence, and slow-moving glacier retreat.

Training data. A labeled dataset used to teach a machine-learning model to recognize patterns. In geospatial applications, training data typically consists of image chips with associated labels (forest, road, ship) that a human analyst has verified. The quality and diversity of training data largely determine how well a model performs.

U

URISA (Urban and Regional Information Systems Association). A professional association focused on the use of spatial information technology in urban planning, public administration, and community development. URISA offers training and publications and co-administers the GISP certification program.

USGIF (United States Geospatial Intelligence Foundation). A nonprofit organization that supports the geospatial intelligence (GEOINT) community through education, workforce development, and the annual GEOINT Symposium. USGIF also administers the GEOINT Professional Certification program.

USGS (United States Geological Survey). A federal science agency within the Department of the Interior. USGS operates the Earth Resources Observation and Science (EROS) Center, manages the Landsat program jointly with NASA, and operates the USGS Earth Explorer data archive where much Landsat, MODIS, and aerial imagery is freely available.

V

Visible light. The narrow slice of the electromagnetic spectrum, roughly 400 to 700 nanometers, that the human eye can detect. Most consumer cameras and many satellite sensors capture red, green, and blue (RGB) visible bands, producing the natural-color images that look familiar to anyone who has used Google Maps.

VIIRS (Visible Infrared Imaging Radiometer Suite). A sensor aboard the Suomi NPP and NOAA-20 satellites that continues and extends the MODIS data record, capturing visible, near-infrared, and thermal-infrared bands at resolutions ranging from 375 to 750 meters. VIIRS is used operationally for fire detection, night-light analysis, and sea-surface temperature mapping.

W

Wildfire (in remote sensing context). Remote sensing has no single "wildfire" term. Still, the concept brings together several glossary entries: thermal infrared detection for active fire fronts, NBR for burn severity mapping, and time-series analysis for monitoring vegetation recovery. Satellite data has become central to real-time wildfire monitoring and post-fire assessment worldwide.

Z

Zenith angle. The angle between the sun and the point directly overhead (the zenith) at a given location and time. Sun zenith angle affects how long sunlight travels through the atmosphere before reaching the ground, influencing how bright a satellite image appears and how much atmospheric correction is needed.

References

Abarca, Victor, et al. "Advances in Multispectral and Hyperspectral Earth Observation for Environmental Monitoring." *Remote Sensing*, vol. 15, no. 4, 2023. (CH2, CH5) Multispectral sensing

Airbus Defence and Space. *Pléiades Neo Technical Specifications*. Airbus, 2024. (CH1, CH3, CH8) Pléiades Neo

Anderson, K., and J. Gaston. "Wildfire Intelligence from Space: Thermal Infrared and SAR Fusion." *International Journal of Wildland Fire*, 2024. (CH1, CH6) Thermal infrared

Bishop, Claire. "AI-Driven Change Detection in High-Resolution Satellite Imagery." *ISPRS Journal of Photogrammetry and Remote Sensing*, 2023. (CH4, CH6, CH9) Change detection

Bourne, Samuel. *The Commercial Earth Observation Market: 2023–2025 Outlook*. Springer, 2023. (CH3, CH8) Earth observation market

Co-Pilot (Word). *Editing and Phrase Suggestions*. Microsoft, 2024. (All Chapters) Co-Pilot Word

Copernicus Programme. *Sentinel-1 SAR User Guide*. European Space Agency, 2024. (CH2, CH6, CH7) Sentinel-1 SAR

Copernicus Programme. *Sentinel-2 MSI Technical Guide*. ESA, 2023. (CH1, CH2, CH5) Sentinel-2 MSI

Deng, Li, et al. "AI-Enhanced Flood Mapping Using SAR and Optical Fusion." *Nature Communications*, 2024. (CH6, CH7, CH9) Flood mapping

DigitalGlobe/Maxar. *WorldView-3 and WorldView-4 Sensor Handbook*. Maxar Technologies, 2023. (CH1, CH3, CH8) WorldView sensors

Feng, Y., and R. Kumar. "Thermal Infrared Remote Sensing for Urban Heat Islands." *Urban Climate*, 2023. (CH2, CH7) Urban heat islands

GEOINT Foundation. *State of GEOINT Report 2024*. USGIF, 2024. (CH8, CH9) GEOINT

Google Gemini. *Image Generation and Editing*. Google, 2024. (All Chapters) Google Gemini

Grammarly AI. *Editing and Spelling Recommendations*. Grammarly, 2024. (All Chapters) Grammarly AI

Huang, Mei, et al. "LiDAR-Enhanced Land Surface Modeling for Precision Agriculture." *Agricultural Systems*, 2023. (CH5) LiDAR

International Charter Space and Major Disasters. *Annual Activity Report 2024*. ICSMD, 2024. (CH1, CH6) Disaster charter

Jensen, John R. *Remote Sensing of the Environment: 2024 Update*. Pearson, 2024. (CH1–CH7) Remote sensing

Khan, A., and S. Patel. "AI-Accelerated SAR Interpretation for Disaster Response." *IEEE Transactions on Geoscience and Remote Sensing*, 2024. (CH6, CH9) SAR interpretation

Lillesand, Thomas, et al. *Remote Sensing and Image Interpretation: 2023 Edition*. Wiley, 2023. (CH1–CH7) Image interpretation

Liu, Z., et al. "Monitoring Deforestation Using Multitemporal Satellite Imagery." *Environmental Research Letters*, 2023. (CH6) Deforestation monitoring

NASA. *Landsat 9 Science Users' Handbook*. NASA Goddard, 2023. (CH1, CH2, CH5) Landsat 9

NASA JPL. *ECOSTRESS Surface Temperature Data Guide*. NASA, 2024. (CH2, CH7) ECOSTRESS

National Academies of Sciences. *Earth Observation from Space: 2023–2025 Priorities*. NAS Press, 2023. (CH3, CH7, CH8) Earth observation priorities

NOAA. *GOES-R Series: Advanced Baseline Imager (ABI) Guide*. NOAA, 2024. (CH2, CH7) GOES ABI

O'Neill, Patrick. "AI-Driven Atmospheric Correction for Satellite Imagery." *Remote Sensing of Environment*, 2024. (CH4, CH7, CH9) Atmospheric correction

Phrasely AI. *Tone Improvement Engine*. Phrasely, 2024. (All Chapters) Phrasely AI

Planet Labs. *PlanetScope and SkySat Imagery Product Specifications*. Planet, 2024. (CH1, CH3, CH5) Planet imagery

Qin, Xiaoyu, et al. "Shortwave Infrared for Soil Moisture and Drought Monitoring." *Geophysical Research Letters*, 2023. (CH2, CH5) SWIR sensing

Ray, Jonathan. "Open-Source Intelligence and the GEOINT Revolution." *Journal of Strategic Security*, 2023. (CH8) OSINT

Sato, Hiroshi, et al. "SAR-Based Infrastructure Damage Assessment After Earthquakes." *Natural Hazards*, 2024. (CH6) Infrastructure damage

Smith, Eleanor. "Thermal Anomalies and Wildfire Spread Modeling." *Fire Ecology*, 2023. (CH1, CH6) Wildfire modeling

UN Food and Agriculture Organization. *Global Crop Monitoring Report 2024*. FAO, 2024. (CH5) Crop monitoring

United Nations Office for Outer Space Affairs. *Space-Based Solutions for Humanitarian Relief 2024*. UNOOSA, 2024. (CH6, CH8) Humanitarian GEOINT

USGS. *3D Elevation Program (3DEP) Annual Report 2024*. USGS, 2024. (CH1, CH5) 3DEP

Wang, Rui, et al. “AI-Enhanced Cloud Removal in Satellite Imagery.” *Machine Learning for Remote Sensing*, 2024. (CH4, CH9) Cloud removal

Zhang, L., and M. Torres. “Time-Series Analysis for Construction Activity Monitoring.” *Remote Sensing Applications*, 2023. (CH6, CH8) Time-series analysis

www.ingramcontent.com/pod-product-compliance
Lightning Source LLC
LaVergne TN
LVHW022203200926
848481LV00006B/1835